A FIELD GUIDE TO CLIMATE ANXIETY

HOW TO **KEEP YOUR COOL** ON A WARMING PLANET

Sarah Jaquette Ray

 UNIVERSITY OF CALIFORNIA PRESS

University of California Press
Oakland, California

© 2020 by Sarah Jaquette Ray

Library of Congress Cataloging-in-Publication Data

Names: Ray, Sarah Jaquette, author.
Title: A field guide to climate anxiety : how to keep your cool on a warming
    planet / Sarah Jaquette Ray.
Description: Oakland, California : University of California Press, [2020] |
    Includes bibliographical references and index.
Identifiers: LCCN 2019049551 (print) | LCCN 2019049552 (ebook) |
    ISBN 9780520343306 (paperback) | ISBN 9780520974722 (ebook)
Subjects: LCSH: Environmental justice. | Environmentalists—Psychology.
Classification: LCC GE230 .R38 2020 (print) | LCC GE230 (ebook) |
    DDC 363.738/74—dc23
LC record available at https://lccn.loc.gov/2019049551
LC ebook record available at https://lccn.loc.gov/2019049552

Manufactured in the United States of America

28 27 26 25 24 23 22
10 9 8 7 6 5 4 3

A FIELD GUIDE TO
CLIMATE ANXIETY

The publisher and the University of California Press Foundation gratefully acknowledge the generous support of the Ralph and Shirley Shapiro Endowment Fund in Environmental Studies.

For Hazel and Daisy, my girls, who first taught me about the preciousness of life and our time on this planet.

Contents

Introduction

Embracing Life in the Anthropocene

Imagine yourself thriving in a climate-changed world. What does your life look like? What needs to happen to you, and what needs to happen around you, to make you feel you have been successful in your efforts to flourish and to improve the lives of others? Imagine ten years from now, being thanked by the next generation for your role in achieving this vision. What exactly are they thanking you for?

Recently, I used these questions in an exercise with my environmental studies college juniors to help them plan their next steps in working toward climate justice. This "vision-change-action" exercise was inspired by community organizer, activist, and professional facilitators Abigail Reyes and adrienne maree brown. I thought my students would love the exercise's concreteness, its centering of their desires and hopes, and its action-oriented, DIY, problem-solving spirit. They often express frustration that their courses are full of information about how bad things are, without giving them ways to tackle those issues. I hoped this exercise would lead them to define the problems themselves and to articulate their own action plans.

In order to figure out what our next steps would be, we needed to imagine the endgame, the ideal state. I asked students to visualize what it would feel and look like to live in a climate-changed future in which all the positive results of all their collective efforts had come to pass. After this visualization exercise, I explained, they would break down the big changes they wanted into doable parts and start strategizing next steps.

This exercise was supposed to be empowering, to free them from the immobilization we all feel in the face of a problem as enormous and intractable as climate change. But it bombed, and in a way I hadn't anticipated. The students *could not visualize a future.* When I asked them about their ideal future state, I heard crickets. When I pushed them to answer, they confessed that they couldn't even form a mental image of the path ahead, much less a future that they could thrive in.

At first, I was confused and even angry with them, mistaking their reaction as short-sightedness or weakness, or an unwillingness to work hard, or a desire to bury their heads in their iPhones. I worried that they were surrendering their agency and using the mounting evidence of apocalypse as an excuse to not roll up their sleeves.

But I soon discovered I was off the mark, and that those who think that "kids these days" are coddled, entitled, and have no grit—that they are "snowflakes"—are simply wrong. The generation growing up in this age of global warming is not lazy or feigning powerlessness. Instead, they are asking *why* they should work hard, and to what end. The bigger problem comes back to their being so frozen by their fears that they are unable to desire—or, yes, even imagine—the future.

The Climate Generation

We live in the Anthropocene—a geological age marked by the irreversible ways in which human beings have affected the climate and environment. Climate change is affecting everyone alive today, and the topics I address and the approaches I offer in what follows are meant for people of all ages. But those of you born between the early 1990s and the early 2000s, at the tail end of the Millennial generation through what is being called Generation Z, or iGen, are the first to have spent your entire lives with the effects of climate change. This is "the climate generation," and I'm addressing you not only because you are uniquely affected by global warming, but also because, more importantly, it is you who are poised to organize and bring about real change.

Your cohort is larger than the Baby Boomers or Generation X. You exist at a time of improved global health, longer lifespans, fewer wars, and greater access to education. But you face a bleaker forecast about the viability of life on this planet than the generations before you. Members of your generation share a mounting awareness that the effects of climate change are not abstract or predicted in some distant future, but are already being felt. The problems on the quickly advancing horizon will diminish the quality of life for everybody.

Your generation shares many characteristics:

- You care greatly about climate change and social justice, and you see a link between the two.
- You feel financially insecure. If you are in college, you will owe on average $40,000 by the time you graduate. If you have graduated, you are worried about finding a job. And you're the first generation that is more likely to be less well off than your parents.
- You're troubled by the increasing disparity between rich and poor.

- You were raised with smart phones, social media, and unprecedented internet access to global networks and information. Although you are more connected, you are also more lonely, suicidal, and depressed than previous generations.
- By the time you reach college, 66 to 85 percent of you have experienced some form of trauma, including sexual assault, violence, loss of a loved one, or bullying.
- Your generation is the most ethnically diverse generation in US history.
- You are the most stressed but also the least likely to vote.

Understanding these distinct qualities of your demographic will better enable you to understand why you're feeling the way you're feeling and how to navigate these turbulent waters.

In 2018, the United Nations Intergovernmental Panel on Climate Change (IPCC) released its fifth report warning of the effects of a 1.5°C (2.7°F) rise in global temperature. The report summarizes the most up-to-date and comprehensive science on climate and the future of the earth from experts around the world, and it also sets targets for emissions reductions. Its 2018 conclusions are far more grim than those of previous assessments. It brings into stark relief that climate change is a fact, not a "belief" that is up for debate, and suggests that the effects of climate change across class and continents will only grow. Climate disruption is no longer something happening "out there" to other people in other places. Sea levels are rising not only in Bangladesh but all along the US coastline. The Mississippi River's banks are overflowing, hurricanes are becoming more extreme along the Gulf and the Eastern Seaboard, and in California the fire season now extends nearly year round. As climate change becomes felt by more people, the boundary between those who worry about a future apocalypse and those who are experiencing that apocalypse right now will further blur. The climate generation is at the cusp of that story.

You are resentful that you are inheriting the problems of previous generations, who have seemingly doomed you to this fate. Earlier generations have reaped the benefits of an extractive, fossil fuel–based economy; they may talk about a looming crisis, but they likely think that serious disruptions will not happen in their own lifetimes. Meanwhile, they may find it difficult to acknowledge their failure to do anything effective about a problem they have seen coming for decades.

Your generation is demanding that climate advocacy attend to social justice as part of any plan for ecological health. Your environmental politics are also eminently *cultural;* unlike your predecessors, you see how important culture and society, not just science and technology, are when it comes to addressing environmental problems.

Yours is the most ethnically diverse generation in US history. For indigenous youth, the issue of climate change has always been connected with colonialism and your own people's stories of genocide and environmental disruption. Many of the rest of you have become politicized around issues such as ocean health, climate refugees, disaster preparation, environmental racism, and the "slow violence" of human suffering that spirals from climate change. Your generation may also be feeling profound despair about the rise of nationalism, xenophobia, and authoritarianism around the globe, together with a lack of progress toward a sustainable and just future. You see daily news cycles of sectarianism, exploitation, and destruction, of which you are often the object. Feelings of grief, mourning, fear, and overwhelm are giving rise to a new vocabulary, including such terms as *climate anxiety, vicarious trauma, solastalgia, pre-traumatic stress,* and *secondary grief,* which are discussed in the next chapter.

A Field Guide to Climate Anxiety argues, however, that a bleak picture is just one side of the story. A shift in climate politics is occurring across the globe, and many are attributing it to Gen Z's burgeoning activism. In the United Kingdom, Extinction Rebellion is a movement of activists who are trying to prevent the extinction of humans by climate change and social collapse. In the United States, the Sunrise Movement is working to bring about a Green New Deal that would regulate emissions at the federal level, even as they focus on organizing local politics toward the same ends. European student climate marches and school strikes inspired by the protests of Swedish student Greta Thunberg suggest that change is coming, and the climate generation is driving it. Along with veteran organizations such as 350.org, the Union of Concerned Scientists, and the Environmental Defense Fund, these groups together constitute what might be termed "the climate movement," an amalgam of many groups that emerged from the broader environmental movement of the 1970s and have taken climate change as their main focus for the current historical moment. By contrast, a combination of younger and social justice groups are reshaping climate conversations and activism around questions of equity, structural violence, systems of power, and identity. Those bridging social justice issues with environmental ones might more likely characterize themselves as a movement for climate *justice.*

Climate change is an on-the-ground issue. The entities that benefit from social injustice are often the same entities that drive climate change, and the polarization of contemporary politics inhibits progress on addressing these issues. Yet the next generation will likely force the Republican Party to change its tune on climate change. Conservatives in Gen Z care about it, while older conservatives do not. It is counterproductive, therefore, to think of climate change as just an issue for Democrats or "liberals," and far more important to begin mobilizing along generational lines.

Your demographic has unique potential: At least 70 percent of your generation cares about global warming, which is about 10 percent more than the general public. You have skills and attitudes that the planet needs: social media savvy to create innovative forms of civic life and community organizing; a lack of faith in existing institutions that will motivate you to reorganize those structures; and awareness of the necessity to face an existential crisis with resilience and solidarity. You are crafting your own emboldening story about yourselves, no longer listening to the stories that older generations are telling about you.

The story you tell about yourself will be crucial in your effort to cope with the changes that are fast coming and to reimagine how the world can be organized. Your story can rise above self-erasure and hatred for humanity, vanquish myths of powerlessness, reject the seduction of denial, and turn away from the distractions of consumable happiness (figure 1). By politicizing your angst, you can focus your energies on collective resilience and adaptation. Making these stories true will require you to nourish, not deny (as much environmentalist doctrine seems to demand) your body and soul. Or, to put it another way: reframing environmentalism as a movement of abundance, connection, and well-being may help us rethink it as a politics of *desire* rather than a politics of individual *sacrifice* and consumer *denial*.

FIGURE 1. "Seashell," by Michael Leunig. Reprinted by permission.

Empowerment and Imagination

Echoing Greta Thunberg, the 1.4 million global youth climate strikers of 2019 were questioning what purpose there was in working hard in school for a future that may never come to pass. They had lost faith in the experts who fail to take the crisis seriously enough. Yet fear and feelings of helplessness and eco-grief (sadness over the destruction of the earth and its species) make it difficult to take long-term action. Such feelings are not good for us or for the planet. Instead, we need to turn them on their heads and tread a path from despair to hope and, crucially, empowerment. For my students, the first step toward progress was the hardest, but as humanities scholar and palliative care expert Marie Eaton explains, we will get nowhere if we do not first "*imagine* the future we hope to live in."

Most of the people disseminating climate messages in the media (80 percent of which are framed in negative terms) are not paying enough attention to the relationship between narrative, emotion, and individual decision-making. That's where environmental humanities comes in. With its focus on storytelling, narrative frames, environmental meaning-making, and ways in which discourse shapes human behavior, this interdisciplinary and wide-ranging field is well situated to help fill this gap.

In this book I look at the research on emotional reactions to climate change and ask, as I asked my students, why throw in the towel? There is a lot of work we can do—problem-solving, building alternative solidarity economies, reducing waste, getting politically organized, making change through our involvement in engineering or politics or law, organizing community, educating. We already have many spheres of influence, and our actions can snowball, opening ever more spheres of impact. How and where and how much to intervene is up to us all individually, and we all need the energy and desire to engage with, not turn away from, the crises we face.

This is not a "how to" guide that will give you practical tips on *how* to do this work—there are already many resources available to help with those kinds of activities. Instead, what you will find here is a focus on actions you can take in how you think about yourself—what I refer to as your "interiority"—drawn from the scholarly knowledge and wisdom I have gained from working with college-aged students for twenty years.

My purpose in writing *A Field Guide to Climate Anxiety* is to support you, the climate generation, and to support the people who work with and care about you. There's no silver bullet that will stop climate change. But we will need a lot of capable and energetic people—i.e., *you*—to navigate the coming storms. This book offers many ways to build and maintain that energy and commitment to climate advocacy, including:

- tools to avoid burnout and to sustain yourselves, your hope, and your community in what trauma worker Laura van Dernoot Lipsky aptly calls "the age of overwhelm"
- techniques to help you find purpose in the climate crisis and to uncover emotional and existential approaches to climate change, sometimes called "adaptive" strategies as opposed to "technical" solutions
- strategies to cultivate personal and collective resilience in the face of the depression, anxiety, fear, and dread that many of us feel when we think about a climate-changed future—or in the face of climate emergencies that are already happening to many of us right now
- tactics to address the doom-and-gloom discussions of climate change in mainstream media
- ways to integrate pleasure, desire, humor, and optimism while pursuing the work of climate justice

Your existential health is the soil in which the future you desire will germinate, as well as the nourishment required to make it a reality.

If the feelings I explore in this book, such as climate anxiety, environmental trauma, and eco-grief, are causing you to feel suicidal or have suicidal ideation, or if you suffer from severe anxiety, depression, or post-traumatic stress disorder related to these issues, you might benefit from professional assistance. This book is not a replacement for psychological support.

In *Emergent Strategy*, adrienne maree brown asks, "How do we cultivate the muscle of radical imagination needed to dream together beyond fear?" As a matter of survival, we need to think beyond eco-apocalypse and nurture our visions for a post–fossil fuel future. Our radical imaginations will also make visible all the good things that are being done, allow us each to see ourselves as a crucial part of a collective movement, and replace the story of a climate-changed future as a frightening battle for ever scarcer resources with one that highlights personal abundance—where there is plenty of time and energy to do the work needed to ensure that we can all be good ancestors to the many generations yet to come.

My Story

Early in my teaching career, I saw my purpose as giving students the intellectual tools to grasp the connections between different kinds of environmental and social problems. I tested students on such concepts as environmental justice, structural violence, and externalities. I asked them to deconstruct their assumptions about nature and the way ideas about nature shape how different communities of people are treated. I taught them how issues of social injustice are tied to environmental issues. They learned about colonialism, capitalism, patriarchy, racism, and how all these -isms can be connected to ecological degradation, and vice versa. I thought my job was to dazzle them with the extent, scale, and scope of interconnected problems, just as environmental science instructors see their mission as disseminating information about how ecosystems work and how human activities impact ecosystem health. Our content may have been different, but our subtext was the same: "if you're not horrified by all of this, then you're not paying attention."

I measured my success by my students' ability to see *more* problems in the world and to question cherished truths. My assumption was that the more students learned about the problems and questioned their own intentions, the better equipped they would be to address environmental problems in a socially just way. Most environmental messages, from documentaries to news reports, are crafted with a similar premise: when audiences can see the problems of the world, they'll be more likely to go out and try to fix them.

I felt I had done my job when I crushed my students' naive, tree-hugging idealism and replaced it with a critical analysis of power. Love wilderness? Don't be a fool! It's socially constructed! Love animals? How childish! You're just projecting your anthropomorphism onto them, when you should really be concerned about the chemicals in your water! Love natural beauty? How elitist! You're just a privileged American! Love science? Don't be duped! It's the product of patriarchal colonial capitalism!

It took me a long time to figure out that these assaults on students' passions were not just creating "discomfort." I had made the students feel guilty about pursuing any kind of pleasure, in nature or otherwise. I hadn't accounted for how their feelings shaped their ability to process the course material, much less do anything about it. I suddenly realized that their emotional responses to learning the extent of the planet's problems, and of their own complicity in those problems, could derail our efforts together. Worse, their despair about the state of the planet and their feelings of guilt (leading to powerlessness) could threaten their ability to show up to class, stay motivated to graduate, and then go into the world with the resolve required to tackle all these problems.

Much to my chagrin, my office hours and classrooms became group therapy sessions, and I found myself totally ill-equipped to

FIGURE 2. The affective arc of environmental studies curricula, by Sarah Jaquette Ray.

manage the demand. Every time a student came to my office, I wondered what crisis, existential or otherwise, I would be expected to sort out this time. The course material was not just a rite of passage, challenging my students to mature into critically thinking adults. It, combined with myriad other stresses of college, was sending my students off the rails, and they were taking me with them.

As I researched emotional reactions to our global crisis, I realized that emotions cannot be separated from what we can think or do about climate change. I began to see patterns in students' emotional responses to the material, which I started to call an "affective arc" of our curriculum (figure 2). I showed students how to reflect on their feelings as stages in an ongoing process and how to understand that this journey was shared by their peers. The vision change workshop I described above, along with the work of adrienne maree brown, led me to think differently about my role as a teacher.

I started to wonder whether it was just as important for students to be asked to address their feelings as it was for them to learn about environmental disruption and injustice. I started to see that instructors needed more tools to help students process the principles of climate disruption in both intellectual and emotional terms. Climate change education rarely accounts for students' emotional lives, even though research shows that emotions are central to our ability to retain and act on information.

Training an emotional lens on climate change debates will enable us to build personal resilience, attend to the social inequities of climate change better, cultivate better relationships and networks, and learn how to talk with a wider range of people, a point I take up in chapter 5. This starts with looking inward, unpacking ourselves, and taking stock of our resources.

The 2016 US presidential election called me to attention. My colleagues in social justice circles were not as shocked as I, and they reminded me that my response was a function of my lack of direct experience with racism and poverty. I knew that I needed more humility; I also needed more tools, and to translate the tools I already taught into more explicit, resilience-building strategies for students. So that's also what this book is about—facing climate grief and injustice with the climate generation, so we can all keep getting up in the morning and doing the work the planet and our community need us to do.

How to Use This Book

Typically, field guides assist with identifying some aspect of the natural world—geology, bird life, tree species, and so forth. They are meant to accompany you in the field, so you can quickly orient yourself and gain richer knowledge of an unfamiliar terrain. They don't tell you everything about a species, but they open the door to intimate knowledge by starting with a name and basic description. This book will lead you through your own interior terrain and help you identify the feelings associated with the environment problems you're working to change. Mindfulness, critical thinking, and emotional intelligence will be your guides on the journey to manifesting the world you desire.

Each chapter explains a strategy for cultivating resilience. The strategies synthesize academic research on emotion, social movement history, affect, and environmental philosophy. They draw on my expertise as an environmental justice and environmental humanities scholar, as well as my experience as a mother, professor, program leader, and certified member of the age of overwhelm.

The chapters need not be read in sequence, nor does the book need to be read in its entirety, though chapters do at times refer to information in other chapters, and many of us struggle with all of the problems the chapters discuss. The various tools may appeal in different ways to different people—or to you differently at different times of your life. Checklists at the end of each chapter distill the key wisdoms. They are refreshers of what the chapter guided you to think about—you might even use them as mantras to orient yourself to the day. The chapter notes at the end of the book describe key concepts in more depth, for further digging and research.

Read through the following chapter summaries, then dive into the chapters that resonate with you.

Chapter 1, "Get Schooled on the Role of Emotions in Climate Justice Work," surveys the research on climate change as it affects mental health. We'll explore the feelings associated with environmental change and how—shaped by class, race, gender, sexuality, power, and identity—they influence our emotional lives. Becoming knowledgeable about these feelings is a first step toward staying cool on a warming planet.

Chapter 2, "Cultivate Climate Wisdom," enlists scholarship about mindfulness, affect theory, grief and trauma, eco-psychology, and emotional intelligence to help us understand the role emotions play in how we think about and work on climate change.

Chapter 3, "Claim Your Calling and Scale Your Action," dismantles two myths that act as barriers to action: (1) the "myth of instrumentalism"—the perception that the results of our actions should always be spectacular and measurable; and (2) the "myth of individualism"—the perception that a single individual cannot make much of a difference.

Chapter 4, "Hack the Story," explores how we can use our imagination to replace stories of urgency and doom with stories of collective and societal transformation. These new stories will help us to stay in climate work for the long haul.

Chapter 5, "Be Less Right and More in Relation," explains how to increase opportunities for collaboration with people from both the right and the left, in part by prioritizing conversation and connection over being right and "winning" debates. Using compassion instead of empathy, we can make social justice and people's material lives more central to climate justice efforts.

Chapter 6, "Move Beyond Hope, Ditch Guilt, and Laugh More," shows guilt, one of the dominant environmentalist emotions, to be destructive and pointless. Rather, as the traditions of "misery resistance" tell us, pleasure, humor, desire, and a critical view of hope are better motivators of long-term commitment. We can find joy in manifesting the world we desire, not just outrage in opposing what we fear.

Chapter 7, "Resist Burnout," describes how to overcome the barriers we put up along our path. Fending off burnout is not simply a narcissistic turn toward interiority at the expense of "the cause" of climate justice; it is in fact essential to the strategy of dismantling existing power relations.

The conclusion, "Feed What You Want to Grow," spells out how the strategies in the book can lead to personal and collective resilience. Rather than focusing all our light and energy on what we fear or what enrages us, we can tend what we want to grow inside ourselves and in the world.

All these chapters help us answer the question, What will it take— *existentially*—to engage in work toward climate justice for the long haul and not just react in fear and panic to today's news of injustice and forecasts of apocalypse?

We need to figure out how to not just *address* but *thrive* in a climate-changed world. *We need to desire, not fear, the future.* I hope this book helps you do that.

1 Get Schooled on the Role of Emotions in Climate Justice Work

One of my students, Madi, wrote a paper about how her environmental values led to such severe self-loathing eco-guilt that she stopped consuming much at all, including food. Taking "zero impact" to its nihilistic endgame, she reduced her footprint by literally reducing the physical space she took up. At a grocery store, when she could not decide on a purchase that would not "somehow contribute to ecological, social, and personal health problems," she would "leave without food altogether, deciding that it was better to go hungry than to make the wrong decision." Madi thought that "to disappear, to become smaller, was to be beautiful"—and, of course, a good environmentalist. Her conclusion was that the best environmentalists are the ones who disappear.

Civil rights attorney and climate activist David Buckel went farther when, in 2018, he immolated himself using gasoline. He claimed in his suicide note that ending his life represented what we're all doing to ourselves by relying on fossil fuels. The first "climate suicide" in the United States, Buckel's death signaled a new intensity in the emotional register of climate change advocacy. Examples of the toll exacted by climate change abound. Photographer Chris Jordan, who spent a year documenting Laysan albatross dying on Midway Atoll from ingesting plastic pollution, has spoken publicly about the depression the experience caused him. For one of my radio journalist colleagues, any mention of Al Gore or climate change is a trigger; he has to turn off the radio when NPR's daily report *Climate Connection* comes on. A viral YouTube video shows a nine-year-old breaking down hysterically when talking about the devastation of the planet.

According to environmental science educator Elin Kelsey, our worries about the environment affect us personally: they "influence what we choose to eat or how we get to work. They keep us awake at night. They make us grieve for the world we are leaving to our grandchildren. They stop us from choosing to have kids. They trigger depression." Yet, she goes on to observe, "there is a strange silence about the emotional impact of the ways in which we talk about the environment." We think nothing of "inviting a scientist into a second-grade classroom and telling the kids the planet is ruined." What's going on here?

Between environmentalism's maxim "leave no trace" and the Church of Euthanasia's creepy slogan "Save the planet, kill yourself," it is no surprise that environmentally conscious Americans suffer from a variety of existential ills, including guilt, depression, grief, and eating disorders. But Buckel's suicide introduced a new, dangerous phase of extreme reaction to environmental disruption. The climate generation is particularly vulnerable to climate anxiety and its attendant ills. Getting informed about the psychological effects specific to climate change, including pre-traumatic stress disorder, solastalgia, and eco-grief, is the first step to overcoming them.

These forms of climate anxiety lead to a feeling of dread about the future combined with a feeling of powerlessness to do anything to shape that future. Climate change disasters have increased rates of suicide, depression, and anxiety in the past and are expected to continue to do so. The new field of disaster mental health has emerged in part to deal with the high psychological stakes of extreme weather events. In the wake of Hurricane Katrina, suicide and suicidal ideation more than doubled; one in six people met the criteria for post-traumatic stress disorder (PTSD), while 49 percent developed an anxiety or mood disorder. Suicide and mood disorders also rise with rising temperatures. Ashlee Cunsolo and Neville R. Ellis, in their article "Ecological Grief as a Mental Health Response to Climate Change–Related Loss," document how "worry about actual or potential impacts of climate change can lead to stress that can build over time," eventually leading to problems such as substance abuse, anxiety disorders, and depression.

Individuals like me who work with people experiencing these symptoms can display additional secondary symptoms such as vicarious trauma, compassion fatigue, or apocalypse fatigue. But these feelings are not limited to people working on the frontlines of trauma, as clinical psychologist Leslie Davenport points out in her book *Emotional Resiliency in an Era of Climate Change.* They are increasingly present in a wider population as more and more people "witness climate-change-induced trauma of loved ones, communities, species, and lands." A whole set of "Anthropocene disorders," as literary critic Timothy Clark calls them in *Ecocriticism on the Edge,* is now part of the mental health and social work vernacular. Although climate anxiety can affect anybody, for some people it can exacerbate other sources of despair or stress, including colonization, military service, displacement, and sexualized violence.

Glenn Albrecht, an Australian philosopher who gives names to the feelings people have in response to their built environments and the earth, calls "earth-related mental health states or conditions" forms of "psychoterratica," or mind-earth states, which can be both pleasant and not. Recognizing that "cultures all over the world have concepts in their languages that relate psychological states to states of the environment," Albrecht found that the English language has few words for environmentally induced distress and illness. He coined the term *solastalgia*—a combination of the Latin word for comfort (*solacium*) and the Greek root for pain (*-algia*)—to capture the existential and psychological feelings people have when their environments undergo profound change or degradation. Natural disasters, land clearing, mining, rapid industrial change, gentrification, terrorism, and war are all possible causes of this solastalgic form of trauma.

Eco-grief can manifest in a number of ways. In "Mourning the Ghost," sustainability activist Amy Spark describes the "anticipatory grief" she felt standing in a forest that was destined to be logged: "The anticipation of knowing the *entirety* of what was going to be lost was much worse than observing the aftermath. It wasn't just trees, but the intrinsic value of the forest and the peace that comes from being there that was going to be destroyed as well." Those who are not yet experiencing climate change's immediate effects can suffer from its indirect effects on agriculture, quality of life, identity (for example, the cultural identity of many Pacific Northwest tribes has been affected by changes in the range and population of salmon), sense of control and autonomy, displacement, and feelings of stress, helplessness, fear, and fatalism. These effects are linked to physical ailments, such as a weakened immune system.

Existential concerns have given birth to an online resource and community of professionals, the Climate Psychology Alliance, whose mission is to "help people face difficult truths." Psychologist

Emily Green describes how thinking about the reality of climate change activates what existential psychology would call our "ultimate concerns" or "existential facts of life, including finitude, responsibility, suffering, meaninglessness, and death." She calls this response "existential dread." Existential Anthropocene disorders affect even people who have been otherwise sheltered from many forms of injustice, suggesting the need for an approach to environmental despair broad enough to encompass people of all races, gender identities, socioeconomic backgrounds, religions, and geographical locations.

An existential fear of the future is entering into the question of whether to have children. In the past, many environmentally minded people decided not to have children in order to reduce population growth; now this decision is coupled with an unwillingness to let one's children live in a state of ongoing ecological crisis. Some young women, Rebecca Solnit observes in *The Mother of All Questions,* choose not to have children because they want to devote their love and energy to climate justice. Congresswoman Alexandria Ocasio-Cortez caused a stir in 2019 when she suggested that it is "legitimate" for women not to have children because of the climate crisis. The new #BirthStrike community is committed to not bearing children "due to the severity of the ecological crisis and the current inaction of governing forces in the face of this existential threat." Given that the climate generation struggles to imagine the future, and given that having children is an act of faith that the future will be desirable, is it any surprise that many are making the choice not to reproduce?

The burgeoning subfield of environmental studies called queer ecological studies is further exploring the relationship between heteronormative nuclear family structures and environmental values. It raises questions like, How does thinking about the future shape sexuality and gender relations? What should we make of women who choose not to reproduce, or people who create new forms of family relationships to build solidarity as economies and social values change in response to climate change?

Environmental educators and sustainability leaders are partly responsible for the attempts of young people like Madi to erase themselves. They (I include myself among them) talk all the time about scarcity, tipping points, and resource limits and insist that we should all moderate our impacts and appetites. They assert that (consumer) desire is the devil to the planet, and that every choice we make has negative consequences. Completing an "ecological footprint" assessment, which shows how many "earths" of resources we use up even if we try to maintain a green lifestyle, is designed to make us feel shocked and guilty about how much we consume. It can have the effect of not just encouraging us to consume less, but of making us think of ourselves—and all of humanity—as bad for the planet.

This attitude is reinforced in school by classes about the environment. "Human Impacts on the Environment" and "Natural Resource Management" evoke a binary that posits humans as greedy consumers and nature as a limited resource for human use. As indigenous students remind me, this framework ignores the contributions of indigenous peoples to ecological health and biodiversity over millennia, creating reciprocal relationships with nature that existed prior to colonialism and modernity and carry on in many communities today. Current indigenous scholarship advances this point further: colonialism and genocide are at the root of our current climate troubles, and the only way to solve them is to return stolen land. Increasingly, as the Green New Deal and Sunrise movements' demands reveal, indigenous sovereignty is core to addressing climate change. To clump all humans into a monolithic category of "bad for the planet" ignores the ways that some humans exploit more resources than others, and the fact that some humans bear the costs of that exploitation more than others. Are all societies equally culpable? Social justice–oriented climate advocacy—or "climate justice"—recognizes that nature is all around us (think of the air we breathe, the water we drink) and that it is possible for humans to actually be good for nature. This contrasts with the dominant view in the United States: that nature is "out there" (not nearby) and, indeed, that in its ideal state it is an untrammeled wilderness that should be protected from humans. This belief directs needed resources and attention away from environmental justice causes. Humans are not separate from nature, which is the infrastructure of all biological life, nor is human activity inherently either "natural" or "unnatural."

The all-humans-are-bad-for-nature narrative does not cultivate the agency that is required to get us up in the morning and get us working on the problems. Instead, it can create a form of self-loathing—which some think is good because it counters the anthropocentrism (a human-centered viewpoint) and humanism (the idea that humans are better than other species) of Judeo-Christian thought. But asking people to see themselves as *only* bad for nature can lead down a rabbit-hole of self-annihilation, as it did Buckel. Framing our environmental crisis in simplistic "humans are bad for the planet" terms undermines our efforts to live better with each other and with nature.

People are profoundly disturbed by climate change, and being told that it is the fault of our own moral failings is not only demoralizing but factually wrong. It does not help us muster the stamina to stay involved in environmental work for the long haul. Instead, it can lead to various forms of self-erasure, or cause people to give up in despair, choosing short-term avoidance and apathy over long-term climate justice.

Social Justice and the Climate Movement

Climate change is inextricable from social justice issues (as we will explore in real-world terms in chapter 5). In the past, however, environmentalists and social justice activists have been somewhat at odds. The climate movement has been perceived as an elitist one, and proposed policies to address climate change have often been seen as a threat by those who fear higher costs for basic necessities or the loss of jobs. Meanwhile, those working for social justice have had cause to be suspicious of Western science, which has sometimes been used to legitimize those in power and their oppressive politics. Western science has at times been manipulated as a form of social control, from the pseudoscience of Social Darwinism to more recent misuses of "science" to delegitimize indigenous or grassroot activists' experiences in places like the tar sands of Alberta and incinerator site locations, where science was constrained by corporate and political interests, so it took a long time to "prove" what was happening. Historically, Western science has also been misused by those in power to "prove" the inferiority of marginalized groups or to supplant their own forms of knowledge.

In contrast, climate justice is often informed by traditional ecological knowledge, arguing that indigenous understanding of the environment is also a kind of science: a lay or "street science." As the more dominant forms of environmental sciences come together with these other ways of understanding the world—what philosophers call "epistemologies"—it becomes difficult to agree even about what the "problem" of climate change is, much less on solutions to that elusive problem. The climate justice movement includes people from a variety of epistemological positions who are making competing claims about what the most pressing problems are and how to solve them. For example, is the root problem colonialism, capitalism, or carbon in the atmosphere? Which should be tackled first? Obviously, they are all pertinent, and intertwined.

Climate justice is an interesting movement in part because it must reconcile the critiques of science with the role science plays in proving that ecological problems exist. This tension is productive, but it causes friction within the broader climate movement, which emphasizes scientific and political solutions that are often experienced outside the daily lives of the people that climate change most affects. Claims to know the "truth" about the environment are complicated by divergent histories, politics, identities, and interests. People in the climate movement don't even share the same definition of nature, let alone other key ideas. Acknowledging multiple epistemologies, multiple subjective experiences, and multiple definitions of "the problem" has resulted in greater awareness of historically marginal perspectives.

Holding competing truths in the same hand can be hard, and it can feel stressful. But this tension gives us an opportunity to understand how important our feelings about truth are to our behaviors and commitments. Confirmation bias—the way we filter information to reinforce our existing viewpoints—and the fact that emotions dictate which facts we care about and which we deem irrelevant show why it's critical to own up to the emotional roots of our climate views and attitudes about the environment.

In subsequent chapters, I will discuss strategies gleaned from psychology, the growing subfield of climate change communication (which teaches about how emotions connect us to environmental problems), and the environmental humanities more broadly. But the insights and resources that have been generated by other social movements, such as those for civil rights, peace, decolonialization and indigenous sovereignty, and women's rights, are a largely untapped asset for the climate movement. Grassroots social justice organizations possess valuable knowledge on how to create and maintain the resilience needed for what environmental justice activist Robert Bullard calls the "marathon" (as opposed to "sprint") of social change. This book seeks to integrate these existential, social justice, and organizing approaches, and in doing so, further celebrate the ways that a social justice focus can improve the climate movement.

What does the climate movement have to gain from the insights of social justice movements? For one, social movements can teach the climate folks techniques of grassroots coalition-building, cultivating collective and personal resilience, understanding the connections between system change and social change, and recognizing the importance of a "heart-hand-head" trilogy in keeping people sustained.

In the climate movement, problem-solving has long been discussed in a top-down, technocratic way, often pushing to the side questions of social justice or efforts to democratize the movement. But since 2014, when the first People's Climate March in New York (with companion actions worldwide) became the largest climate march in history and introduced climate justice as a mobilizing concept, the movement has undergone a significant shift toward addressing social justice—resulting in the growth of the climate *justice* movement. One outcome of this development is a greater focus on the cultural, social, organizing, and emotional dimensions of climate work, whether that work is in science, technology, politics, community organizing, creative expression, teaching, or activism.

Evidence is growing that climate change has a disproportionate impact on the poor. Environmental forms of trauma such as solastalgia will exacerbate existing forms of oppression. Insights from trauma studies and social movements such as the Movement for Black Lives (M4BL), in addition to traditions emphasizing resilience in a variety of movements for social justice, from indigenous to women's to civil rights, reveal that personal healing and advocating for social change are two sides of the same coin. Ensuring that we are emotionally and existentially resourced is not a navel-gazing privilege; on the contrary, as movement strategists know, the interior work that healing from oppression requires is exactly the kind of work that enables us to be social change agents. Anybody who wants to work on climate justice for the long haul will need these resilience-building skills in order to do the external work of climate advocacy and community engagement, a point I take up in chapter 7.

The key to happiness is not pleasure or the absence of discomfort, but a sense of purpose. Emotional intelligence, self-regulation, understanding our "windows of tolerance," mindfulness, and critical thinking, which I turn to in the next chapter, are all strategies for sustaining that sense of purpose. Cultivating existential resilience is a *necessity for*—not a *distraction from*—the work of changing the social structures that create these myriad forms of interlinked oppression.

Learning How to Identify Climate Anxiety

☐ Environmentally related mental health issues exist and are pervasive; acceptance of this is the first step.

☐ Self-hatred and misanthropy are not useful, or sustainable, if you want to effect change.

☐ Climate anxiety is diverse: know how gender, class, race, age, ability, and location shape your outlook.

☐ Resilience is critical to the movement, not a luxury outside of it.

2   Cultivate Climate Wisdom

Climate wisdom is the understanding that our ability to respond to climate change and to work on climate issues is shaped more by our emotional selves than by our rational selves. The sooner we can make that connection, the more effective we will be.

One day, my student Job unraveled in a tirade about the futility of everything we were doing and learning. Looking at the ceiling, leaning back in frustration, he called all of our efforts "failures" and railed for over ten minutes about how all of our discussions, our projects, and our classes were a waste of time. They weren't even *touching* the huge problem of climate change. His intensity silenced and shamed us. Those of us who felt we *were* finding purpose in the class slunk down in our chairs, as if his words exposed us as deluded. People who had struggled even to get to college wondered if it was just a waste of money and their family's sacrifices.

This was one of the lowest moments in my teaching life. Job was doing more than just "killing our buzz"; he was tearing at the delicate strings that were holding us together and helping us get to class every day. I have thought about that experience often, and each time I've gained more from it.

Job rejected our response that "love" for the planet could be an antidote to despair over its degradation, and instead argued that love was a joke; anger was more appropriate and valuable. Yet had he known more about the intimate relationship between rage and love—that anger is a response to the fear of losing that which you love—he would have seen that his anger was *part of* love, not a rejection of it. Had I been trained in emotional intelligence as a professor, I could have more tactfully enlisted his feelings in support of our collective efforts, rather than let them deflate our hard-earned and fragile sense of agency. What I now understand is that Job was going through what I call the "nihilist" stage of the climate grieving process. Instead of allowing him to work through it, we took his attacks personally and retreated, thinking that the classroom was not a place for emotions. I have since learned how to handle these incidents as opportunities to build community, explore the complex emotional terrain of climate action, and identify strategies that help everyone feel more successful.

Emotional skills are as crucial for advancing climate justice as technical or political skills. This chapter combines insights from a variety of fields to explain why it's important to take emotions about climate change seriously. I draw on affect theory, secular mindfulness, eco-psychology, and critical thinking, as well as the organizing work of adrienne maree brown and Buddhist eco-psychologist Joanna Macy, to show how to effectively center your emotions. Attending to the emotional challenges of climate work will not only help the already converted avoid emotional exhaustion, it will also make climate change feel relevant and compelling for some people who are not already engaged.

Science and Emotion

Climate scientists tend to put the whims of emotion to the side. To generate trust, they refrain from demonstrating human bias or subjectivity and strive to stay not only emotionally detached from the environmental crises they study, but also politically neutral. If they start engaging in culture wars, they can undermine the very environment they hope to save.

But they have political viewpoints and emotions too. In fact, environmental scientists suffer some of the highest rates of eco-despair—the feeling of grief about what's happening to the planet—being in the unenviable position of witnessing not just the loss of biodiversity and natural wonders but also the loss of their life's work. This includes biologists studying threatened and endangered species like whooping cranes and rhinoceroses; earth scientists recording the accelerated destruction of natural features such as coral reefs and glaciers; and all manner of other specialists—ornithologists, herpetologists, ichthyologists, botanists—watching complex ecosystems shrink or disappear. While objectivity may be a scientific ideal, it takes a huge emotional toll to watch what you care about vanish. This despair, however, does not stem just from sentimentality. Environmental scientists are acutely aware of each species' role in their ecosystems, and warn that the ecological disruption caused by habitat and species loss will eventually impact our ability to sustain ourselves.

The charge that climate change is a fraud perpetrated by the scientific community has made many scientists feel they must enter the political world, despite the threat this poses to their credibility as nonpartisan and objective fact-finders and the potential cost of their careers. They feel pressure to "march for science" and to defend science against "alternative facts" and "fake news."

Entering the fray can backfire, because when progressives proclaim science as a liberal value, conservatives are more apt to reject it, thus closing the door on conversations that need to happen.

Many scientists are being more open about their emotional investment, encouraged by the fact that the social sciences are proving (using science, of course) that emotions are central to engaging the public and inspiring action. However, their professional focus remains on providing information about climate change; they do not address the emotional responses people have to that information, or the emotional tools needed for the organizing, political, and cultural work that climate change demands of us. And that is not enough. The way ahead will require not just the science of *climate*, but the science of *emotions*, to help us balance apathy, fear, and despair with efficacy, compassion, and desire.

This imbalance is just one dimension of a larger problem, which is that Western culture in general has underprioritized the importance of emotion in daily life. "Emotional intelligence" is a term coined in 1964 and popularized in the eponymous 1995 book by Daniel Goleman. It has been taken up as an approach in a variety of areas, such as education and child development. The term combines two concepts that are often regarded as at odds with each other: to be "emotional" has been seen as having nothing to do with "intelligence," especially when race, gender, and class are factored in. (Every female-identified person knows she's being insulted as unintelligent when she's called "too emotional," and male-identified folks might take such an accusation as an insult to their masculinity.) In our Western culture, which prioritizes reason over feeling, emotional intelligence is a radical concept.

Environmental educators Pamela Barker and Amy McConnell Franklin define emotional intelligence as the capacity of a person to "use emotion to enhance reasoning and decision-making." The more we focus on emotions, the more we learn to draw on the ones that serve us and recognize the ones that do not. We also discover how to avoid being manipulated by the messages we receive from sources that would benefit from destabilizing our emotional grounding. Insecurity, fear, loneliness—when we feel these, we are susceptible to being controlled by others. Paradoxically, closely observing our emotions actually makes us behave more rationally.

In this seriously challenging time, we all need the emotional skills to face traumatic information, loss, and instability. Let's bring to the job the tools it requires, and stop pretending the issue is merely a battle between facts and alternative facts.

Less Information, More Feelings

I was once asked to address a group of first-year college students in a joint talk with a climate scientist. He offered to speak first so I could follow his "science" with what he thought I was an expert in—the "human impacts" of climate change. His assumption about what I study revealed his prejudice that the scientific facts should come first, and then the rest of us should figure out what to do with them, that the only insights humanists could offer was to describe how people are dealing with the warming climate. His slides were redolent with images of icecaps melting, jagged-lined charts, and boldfaced headings such as "Go to the Data!" and "What Are the Facts?" He thought his job was to stridently persuade the students of the problems represented by the red blobs of heat moving over the planet in his animated slides. In my presentation, I spoke about how humanities scholars care not just about how climate change will affect people, but about how the *narratives* of climate change affect people. I explained that humans *produce* science, they don't just *react to* scientific facts, and so we ought to think about who gets to determine which facts to pursue as much as what they mean for us. I also offered the point that even if scientists could provide certainty, facts themselves have much less bearing on human decision-making than my colleague obviously believed. But this is how a lot of climate scientists think they should be talking to people.

Within these cries of doom is a common trope: a litany of "proof" that climate change is happening, it's real, and it's going to be awful for all of us, even the privileged. Simply listing the horrors, however, in what environmental literary theorist Heather Houser calls "infowhelm," is a poor response to the so-called debate in the dominant media about whether or not climate change is happening. Writers, thinkers, and leaders feel the need to bombard us with the specter of apocalypse, citing scientific facts that they think we don't already know. But many of us have the famous "hockey stick" graph of rising atmospheric carbon emblazoned on our psyches. Grief for extinct species, threatened cultures, and climate refugees haunts our sleep. At best, the sledgehammer approach is not relevant to those who already care a lot, and at worst, it is destructive to our efforts to enlist more people on our side. The science of emotion tells us that such litanies prompt passivity, not action. Doomsayers can be as much a problem for the climate movement as deniers, because they spark guilt, fear, apathy, nihilism, and ultimately inertia. Who wants to join that movement?

In *Living in Denial: Climate Change, Emotions, and Everyday Life*, sociologist Kari Norgaard examines a variety of communities' responses to climate change information. Her data show that the climate movement operates according to a myth of "information deficit"—the idea that, in order to make a decision to act on climate change, people need ever *more* information and evidence. But this is not how things work. Scott and Paul Slovic, in their introduction to *Numbers and Nerves: Information, Emotion, and Meaning,* argue that using numbers and data to get people to care about and act on climate change can "leave audiences numb and messages devoid of meaning." Instead, Norwegian psychologist Per Espen Stoknes argues, how we feel about the *messenger* of the information is more likely to shape our decision to act or not than the message itself. We are more likely to accept information and solutions when they come from someone whom we perceive as sharing our core values. For example, a conservative person who enjoys hunting or other outdoor recreation is likely to be concerned about climate change. However, they are also more likely to be skeptical about climate change information and advice on climate policy if it comes from Al Gore. Alex Bozmoski, co-founder of the national organization RepublicEn (their tagline: "Energy Optimists— Climate Realists"), observed, "When you don't trust anyone talking about climate change, when you don't see your tribe talking about solutions that fit with your worldview, it's really easy to cope with the problem by ignoring it or denying it." RepublicEn's mission is to garner conservative support for a carbon tax by showing how lowering greenhouse gas emissions goes hand in hand with conservative values. Here we see that the identity of the messenger may be a bigger factor in persuading people than scientific certainty.

In *What We Think about When We Try Not to Think about Global Warming,* Per Espen Stoknes extends Norgaard's ideas about denial and identity, and shows that the more we learn about climate change, the more we resist doing anything about it. He suggests that there are five main psychological barriers that lead to climate inaction, all tied up in an emotional rather than a rational

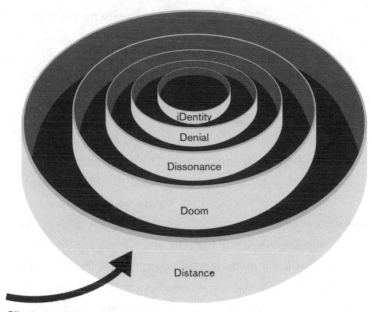

iDentity

Denial

Dissonance

Doom

Distance

Climate message

FIGURE 3. The 5 Ds of climate communication. Reprinted from *What We Think about When We Try Not to Think about Global Warming*, copyright 2015 by Per Espen Stoknes, with permission from Chelsea Green Publishing.

response—Distance, Doom, Dissonance, Denial, and iDentity (figure 3). Distance is the perception that climate change is both geographically and temporally far away, while doom is the aversion we feel to news that the world as we know it is coming to an end. Denial and dissonance are intimately connected: we feel dissonance when we experience two feelings that are at odds (e.g., guilt and powerlessness), which often leads to denial about the problems. Importantly, Stoknes argues that more information actually *deflates* our ability to muster a response. Finally, one's identification with the messenger of climate information (iDentity, in

Stoknes's framework) significantly affects how a person interprets and acts on that information.

Mainstream discussions about climate change would have us believe that there are only two ways to view climate change: you're either a believer, like Bill McKibben, who has been leading the climate charge for over thirty years, or a denier, like Donald Trump, who claims that climate change is a "Chinese hoax." But the reality is that there's a wide range of stances. The Yale Program on Climate Change Communication, in an article titled "Global Warming's Six Americas," describes not two but six groups that Americans fall into with respect to their perceptions of climate change. They range from the Alarmed, who are fully engaged; through the Concerned, the Cautious, the Disengaged, and the Doubtful; to the Dismissive, who oppose all efforts to reduce greenhouse gas emissions. Thus, contrary to what social media articles, news outlets, friends, classes, and parents tell us, it's not a battle between deniers and believers, or those who are wrong and those who are right. Rethinking who the "enemy" is can change our emotional orientation toward climate action. There's no monolithic army of hostile opponents, but rather a fragmented group of stakeholders with disparate interests, understandings, and needs.

How can we approach the "enemy," in all its variety? Moving away from abstractions like "atmosphere," "Anthropocene," and even ice caps to talking instead about daily struggles for justice, like infrastructure and access to environmental goods, goes a long way toward making climate change emotionally accessible to a range of people. This is a crucial insight that I'll take up in more detail in chapter 5. There are many reasons why people care about climate change, and many reasons why they don't. Some have to do with identity, such as class or race, and some have to do with the challenges of assigning a coherent set of emotions to such an amorphous thing as climate change. It's a wide field.

Let's start with affect theory.

How Does Our Affect Affect Our Effectiveness?

Affect theory—the study of emotion in fields such as philosophy, queer studies, and cultural and literary studies—helps us understand how our attachment to happiness shapes how we cope in life. Happiness is particularly prized in American culture. The "pursuit of happiness," written into the Declaration of Independence, is considered a right of American citizens. Happiness has come to stand in for the utopic goal of human potential and the dream of liberal democracy. Yet Americans are not as happy as you would expect, given the nation's current prosperity and positive trends such as lower crime rates, lower death rates from cancer, and greater literacy. Is happiness getting in the way of our happiness?

Affect theorists interrogate the notion of happiness by asking how definitions have changed over time and whether it's helpful to build an entire nation on the premise that citizenship entails the right to pursue happiness. The ideal of happiness may, ironically, undermine resilience and well-being in the pursuit of justice of all kinds. That is, good feelings like joy exist alongside grave inequality, while bad feelings like despair are not inconsistent with positive developments in the world. In the case of climate justice, it may give us happiness to eat meat or to explore distant parts of the world, even though doing so may cause the suffering of invisible, seemingly disposable people and environments. Coming to terms with human suffering may cause unhappiness, but awareness is the first step toward living justly. Realizing that our good and bad feelings are not always aligned with whether things are going well for others is crucial to building resilience. We can be doing good and feel great purpose and yet not be happy. In fact, suppressing our "bad" feelings, such as despair, sadness, and mourning, can lead to what psychologists call avoidance behaviors and emotional dysregulation (as opposed to self-regulation, which is the ideal). Similarly, we can be doing great damage to the planet but remain perfectly happy if we're ignorant of the consequences of our actions. Coping with the enormous threat of climate change requires that we challenge our attachment to happiness and our aversion to discomfort.

If we expect happiness to be the default, we are less likely to be comfortable with negative feelings. Inability to grapple with negative affects can lead to denial, or worse, what I call eco-nihilism—the notion that we should just erase ourselves because we are so bad for the planet.

On the other hand, some people thrive in "the struggle" and feel like there's something wrong if they start to feel happy and well adjusted, as if this shows that they do not care enough about the cause. Many of us feel we are complicit in causing harm if we do not suffer alongside those who suffer the most.

Although this tendency comes from a generous place of wanting to be present with those who suffer, our perception that the crisis is urgent and requires immediate and intense engagement can lead to burnout. If we really care about a cause, our priority should be to avoid martyrdom. Despite what some of us have internalized from the environmental movement, our presence on the planet does not have to mean eternal damnation. We are all worthy of living on Earth, and we do not need to spend our lives repenting the sin of being human. (As I discuss more in chapter 6, guilt isn't an effective affect for climate justice—or much else, for that matter.)

The ethos of sacrifice that underwrites so much environmental work is not a path toward effectiveness; on the contrary, it is *disabling*. Organizer and writer Marisol Cortez, for one, bemoans the activist sense of urgency that results in constantly chasing after fires. She tells her story of recovering from activist burnout:

> The organizational culture of social justice nonprofits is deeply ableist in its productivism, internalizing self-injuring narratives about the valor of exhaustion and working beyond one's capacity, both as an individual and as an organization. So much of community work, not unlike academic work, is grounded in assumptions that we are autonomous—wholly free to give our lives wholly, as though we were not also embedded in social ecologies of interdependence, responsible and responsive to the needs of others. For those who live with chronic mental or physical illness, or for those who do caregiving work, the assumption that our bodies and minds can sustain constant conflict, constant confrontation, constant crisis in the name of justice or sustainability is a disabling one. Eventually I too got sick and had to leave my job as a paid organizer and community-based scholar for my own survival.

One of my students, Gabi, confessed that her solution to juggling work, academics, and activism was to cut down on sleep and to cope with her feeling of overwhelm by drinking too much beer. Such busyness and unhealthfulness had become a marker of her commitment to being a social change agent. These badges of honor, however, make sense only within the "productivist" worldview that is causing ecological collapse in the first place. To burn out trying to resist a system that is fueled by burning things out is not resistance—a point I take up in more detail in chapter 7. Neither *overattachment to* nor *total suspicion of* happiness is good for the climate generation or, by extension, the climate.

In chapter 4, I explore why the mode of urgency has become the unhealthy currency of so much environmental discourse. It motivates people to feel they need to sacrifice their own comfort and health in order to save us all from climate injustice. But for now, I want to stress that happiness isn't the point, and neither is its opposite—suffering. When these become the *modus operandi* of climate advocates, they only reinforce isolation and exhaustion. Cortez urges us to "decelerate" and pay attention to the *how,* not just the *what,* of activism in a kind of slow "praxis," or theory of practice. How can we measure our success by something other than capitalist-defined qualities, like our deliverables or our quarterly returns or our "happiness"? How do we shift attention from the outcomes toward the process? Certainly outcomes are important—feeling efficacious can help us to stay in the game—but overemphasis on results can lead people to abandon it. The dangers of frenzied resistance outweigh the benefits.

What Mindfulness Can Bring to the Climate Movement

Mindfulness, a practice of staying in the moment, is one of the key forms of emotional intelligence. Originally used by Brahmans and later by Buddhist monks as a way to attain enlightenment, mindfulness is increasingly appreciated by neuroscientists for its ability to enhance self-regulation and foster a sense of agency. Shunryu Suzuki's *Zen Mind, Beginner's Mind: Informal Talks on Meditation and Practice* helped popularize practices of meditation and observing the effects of our minds on our actions. But mindfulness has been taken up in secular ways too, as in the work of Jon Kabat-Zinn, who developed the Mindfulness-Based Stress Reduction program. He defines it as "paying attention in a particular way: on purpose, in the present moment, and non-judgmentally." Mindfulness helps us be less reactive to all of the signals we receive from the world and from others. It helps us recognize the roots of our emotions, and to bring those roots "above the line," as Tara Brach puts it, where we can see them for what they are and then think about how to deal with them. Meditation is one example of a mindfulness practice, but so too is any effort to observe our minds, bodies, and feelings in any given moment. Writer and educator David Treleaven, who works at the intersection of mindfulness and trauma, describes the process this way: "By virtue of paying close, nonjudgmental attention to their inner world, people who practice mindfulness are more self-responsive to their own emotions and can even have less emotion exhaustion. [Mindfulness] also increases their capacity to be present with challenging emotions and thoughts without overreacting."

Reckoning with the impermanence of our own lives, and possibly much of the life on Earth, can bring on existential grief. Through mindfulness, we can see that the root of that grief is fear of loss or the terror of loneliness. We can assuage those root feelings by tending relationships or appreciating the beauty and moments we *do* have. These are far more effective ways to address existential grief than extinguishing ourselves under the weight of it all. Accepting the negative aspects of climate change can open the door for positive feelings that will last longer than the hit we get from clicking "like" on our grim newsfeed. In this way, the self-help practices of stopping at intervals to take three grounding deep breaths or of listing five things we are grateful for at the end of each day are not just ways to focus on the positive or to see the glass as half full so we can feel better; crucially, mindfulness is a praxis of *environmental* (not just *self*) preservation.

Mindfulness also opens our eyes to the realization that the pursuit of happiness distracts us from understanding what is meaningful in our lives. We don't need to experience unpleasant emotions all the time, wallow in them, or wear them as a badge of legitimacy, but we should recognize that grief, despair, anxiety, and rage all arise from love, desire, and compassion. We cannot have the latter without the former.

Buddhism supports affect theory in shifting attention away from happiness or martyrdom as worthwhile goals. Both Buddhism and affect theory counsel that we should not be attached to happiness, and also that asceticism is its own kind of illusion. Buddhism's first noble truth is, after all, "all life is suffering." As an undergraduate in a religious studies major, I used to think this was a dour way to look at things. I also didn't like that Buddhism rejected desire, viewing it as the root of suffering. It seemed to me that desire led to pleasure. But I have come to understand that this first noble truth is not *prescriptive*—it's not that all life *should* be about suffering. Rather, it is *descriptive:* when humans grasp for relief from the discomforts of life and try to escape the existential knowledge of their own and everyone else's imminent death, suffering happens. The grasping we do to dodge this discomfort causes more suffering—it's a feedback loop. The opposite of suffering is not happiness; it's compassionate acceptance of negative feelings. Our reptilian brain's mindless aversion to negative feelings and our striving for relief of suffering only cause more suffering. Aversion and grasping are just Band-Aids on the wound. When we operate in this dance of aversion/grasping, we exhaust ourselves and never wake up from what Brach calls the "trance," the illusion that we are supposed to be happy all the time. Happiness obtained by avoiding discomfort and seeking pleasures is addictive, yet ultimately fleeting, like all addictive highs. Accepting that death happens, like accepting that climate change is calamitous, is a door to equanimity *despite* suffering, and leads to that elusive existential holy grail—resilience (which I contend is the solution to anguish about climate change).

The second noble truth is that we suffer because we desire, and we grasp at things that we think will give us happiness or relief from those discomforts. To be honest, I like the feeling of being desirous, and I believe that desire is more effective for sustaining our engagement in the world than happiness, so this one is also hard for me to square with my hedonistic values. But Buddhism distinguishes between the kinds of desires that are distractions from true fulfillment, on the one hand, and the desire for wisdom and the alleviation of suffering, on the other. The latter is good; the former creates stress.

The trick is to choose between the quick dopamine hits of pleasure, like the bottle of wine I drink when I want to stop feeling despair about the suffering in the world that I can do nothing to fix, and the desire that is about deeper acceptance of the purpose of our existence. I don't really desire a bottle of wine, in other words; I actually desire relief from suffering, my own and everything else's. Sometimes pursuing this latter form of desire can paradoxically feel like a kind of ascetic denial, like going on a fast, as we turn away from self-sabotaging habits that may feel good in the short term, in favor of tending to our deeper desires for the long term. It's crucial to distinguish between these kinds of desire and pleasure. The dopamine-hit distractions lead to precisely the problems that are causing the climate crisis in the first place. How many times have I opened my Amazon app to buy something meaningless instead of taken time to pause, observe, and investigate my twitchy feeling of angst about some bad news in the world? Relief of suffering starts with self-compassion. In facing and embracing my discomfort rather than trying to escape it, I can reduce further harm (impulse-buying on Amazon) in favor of a deliberate action that addresses my root feelings (taking a walk). When I choose the latter, I can transform myself, but because this choice also reduces consumption and increases my energy for engagement, it has a broader impact on the world.

The third and fourth noble truths in Buddhism posit that we can alleviate suffering not through attaining that which we grasp after, like more stuff or distractions, but by releasing ourselves from grasping in the first place (through what Buddhists call the Eightfold Path—right view, right intention, right speech, right conduct, right livelihood, right effort, right mindfulness, and right insight). Becoming more aware of our feelings in response to suffering in the world, including our own, allows us to experience rather than sidestep that suffering. The reason we need to experience suffering is that, ironically, trying to escape it is likely to take a toll on the well-being of others, ourselves, and the planet, perpetuating a cycle of suffering.

When we are able to acknowledge and investigate our own experiences of suffering, we are no longer driven by aversion to pain itself. Instead, we can self-regulate by observing the force that pain exerts on us—depleting our energy, repelling us—and then making decisions about how to act in spite (or because) of it. Magnetic resonance imaging (MRI) technology research on the brain shows why mindfulness, emotional intelligence, and other forms of self-regulation help. Emotions like anger and fear reduce activity in the parts of the brain associated with reason, diminishing our ability to modulate emotional responses. The limbic system—responsible for activities we can do when we are born, like eat, breathe, and urinate—is activated by strong emotions, which, for good evolutionary reasons, bypass logic to help us fight, flee, or freeze. The prefrontal cortex helps us interpret information and assess threats. In this brain research, mindfulness practices show greater activation in the prefrontal cortex and decreased volume of gray matter in the amygdala, which makes judgments before the rational part of the brain can weigh in.

Brach recommends that when we feel threatened or over-whelmed, we imagine ourselves *as* the waves in the ocean, rather than feeling pummeled *by* the waves. We can do this by disrupting our limbic system reaction, which triggers our fear instincts, and instead practicing what she calls RAIN: *Recognize* what is happening, *Allow* our feeling to be what it is, *Investigate* it with gentle attention, and *Nurture* ourselves with self-compassion. The RAIN process gets to our root fears by spending nonjudgmental time with them and being compassionate toward ourselves for having these emotions. In Buddhism, self-regulation is built into the Eightfold Path. In affect theory, becoming aware of the work that emotions do in our culture helps us avoid being manipulated by the narratives we consume daily, and in turn makes us conscious of the impacts of the stories we tell. All of these approaches help us navigate the feedback loop between our interior lives and the external world. They put emotions at the center, even as they teach us to act from those emotions in effective rather than reactionary ways. As emotional, spiritual, and intellectual tools, they allow us to select from a range of options without dogmatically subscribing to any one in particular.

All "negative" emotions—including grief, anger, and anxiety—have a role to play in our lives. The key to managing them is to find equanimity from which to act effectively, so that we are not reacting endlessly to all of the upsetting stimuli around us. Allowing, observing, and investigating a wide palette of feelings helps us avoid the pitfalls of burnout, fatalism, or denial. Our feelings become entities that we can talk about and look straight in the eye, rather than run away from. I often imagine myself sitting on a park bench next to my despair. I feel its presence next to me, and I ask, Where are you coming from? What do I fear most? What good can come of this?

My existential tool kit would not be complete without critical thinking. Because I'm not a trained psychologist or ordained Buddhist teacher, I approach these phenomena intellectually and try to draw on what I was trained to do as a scholar—use the humanities methodology of analyzing texts and messages in politics and culture. The scholarly tools of the humanities, such as close reading, discourse analysis, historical perspective, and creative expression, can help with emotional self-regulation. These instruments can help us analyze how different ideas take hold of our emotions. Critical thinking promotes the cultivation of a radical imagination *about* and desire *for* the future, even if some explorations—such as deconstructing existing power relations, seeing sources of exploitation as interconnected, and examining the historical contexts of our identities—might make us feel "bad." Social and civil rights movements have long known that the heart, the hand, and the head are all required for sustained engagement in the movement.

I invite you to practice this methodology in your own life. In a journal or with friends, try recognizing the affective implications of the content you are exposed to, and ask, Which narratives have made us feel this way? Are there other ways to feel? Who benefits from us feeling this way? Who loses? How do our feelings shape our ability to engage in civic life? What feelings do we need in order to sustain sense of purpose for the long haul of advocating for climate justice?

Begin with Beginner's Mind

Becoming intimately aware of one's feelings is the first step to not letting them whip us about with every message we consume and every encounter we have. The point of emotional intelligence is to understand how our emotions operate in our lives to serve (or not) our well-being and interests. Studies on mindfulness and emotional intelligence tell us that the ability to stand back and observe our emotions, identify core beliefs from which emotions emerge, and disrupt the limbic fight-or-flight response is the key to dealing with emotions.

Being mindful about the stories we consume and how they make us feel isn't about indulging our emotional responses by doing the most immediate thing we want to do—rage on Twitter, join a protest, scream the loudest, or do nothing. Rather, it's about identifying the most efficacious action for *you* given the problem and your relationship to it. We need to conserve energy; we are of no use if we burn out. Being mindful about where to devote our time and energy is not a way to get out of the work. On the contrary, it is a way to preserve ourselves for the long haul by not flapping with every wind that blows. It is a way to strategize where best to devote our limited energies.

For each person, the solution will be very different. Behavioral change experts suggest that we start by breaking apart our overwhelming concerns into small steps toward which we can "nudge" ourselves. Working backward from the huge problem, what is the very first step, no matter how seemingly small, we can each take?

One of my students, Angel, answered this question by printing out a mindfulness worksheet and doing things like brushing her teeth with her nondominant hand. Does this count as an effective action that will address Angel's greatest concerns? Not likely. What Angel was saying to herself was, "The world's problems overwhelm me. I don't know where to begin. I feel powerless. How will I make a difference? Where will I begin?" She was cultivating the open-minded curiosity that Zen practitioners call "beginner's mind." The beginner's mind approaches every moment and every challenge with a sense of possibility, curiosity, and humility. In Angel's case, it helped her clear out the clutter in order to build her purpose from the ground up. Why do we need beginner's mind right now? Because in order to think critically outside of binary political sides, and in order to cultivate a radical imagination about a future we desire, we have to get out of our comfort zones. How do you nudge yourself in these really important directions? Observe yourself. Beginner's mind.

UPPING YOUR EMOTIONAL RESILIENCE MEANS ▶▶

Focusing First on What You Can Control

☐ You have more options than fight/flight/freeze. You can regulate your feelings.

☐ You can harness mindfulness as praxis.

☐ You can survive discomfort and use it for good.

☐ You don't need to be "always on." You should *not* be "always on." That's what got us here in the first place.

- ☐ Contextualize and interrogate "happiness."

- ☐ Reject reason/feeling, mind/heart dualisms in favor of productive gray areas.

- ☐ Have compassion for your negative feelings.

- ☐ Favor nudging over nihilism.

3   Claim Your Calling and Scale Your Action

Ron Finley, the self-described "gangsta gardener" from South Central Los Angeles, did not set out to start a revolution. "I wanted a carrot without toxic ingredients I didn't know how to spell," he says. Tired of driving long distances for healthy food, in 2010 he started planting vegetables in the curbside strip of dirt near his home. The city of Los Angeles issued an arrest warrant, citing the garden as an illegal use of public space. Working with fellow activists, Ron successfully fought back, demanding the right to grow food in the city. He continues to empower residents in the neighborhood to grow their own food and has established the Ron Finley Project to build an urban garden in South Central LA to serve "as an example of a well-balanced fruit and veggie oasis." He is asked to speak around the world about the importance of urban gardens to community and environmental health.

Finley's story is an example of the power we can all find when we identify our spheres of influence, redefine action, find strength in numbers, and cultivate emotional resilience. Our sense of purpose will sustain us in our climate work.

Two assumptions can thwart these goals. One is the belief that you must have a lot of power, in the sense of having elevated social, economic, or political status—the illusion that politicians, religious leaders, or celebrities are the only people whose efforts can make a difference. The second is that your actions need to be measurably impactful and yield immediate results—an expectation that can make you unhappy with your efforts if you don't see change happening.

These assumptions are barriers to resilience. The holy grail of fulfillment is not "happiness," but finding our purpose or individual calling in life, or *dharma*. We need to discover where our personal power comes from, generate confidence that our individual efforts will make a difference, and explore our role in working with others who are also pushing the "leverage points" within their "spheres of influence"—all concepts I take up in this chapter. We also need to think broadly about what change looks like and how we can think about it at a human scale.

Our realms of adaptation activism will vary widely depending on who we are. As Dahr Jamail and Barbara Cecil, in their exploration of activism in the context of environmental collapse, remind us, "Each one of us must choose the path that is ours. The sum total of this is legions of people taking action in their unique ways, and supporting one another." Setting a course starts with knowing as precisely as possible which direction you want to go to reach your destination. The earlier you set the right course, the less energy you'll waste.

Ask yourself what kind of work, personal and public, gives you pleasure. Once you start down your path, the positive feelings you need in order to carry on will come. The despair, grief, and anxiety may never go away, but they'll have good company with hope, agency, and resilience. You don't need to feel the good emotions first, and only then act. As Buddhist logic has it, "Right thinking follows right action," not the other way around. Find ways to engage that honor your true capacities and passions, rather than some standard defined by the IPCC, your family, your best friend, or Leonardo DiCaprio.

Social movement theory holds that change happens on myriad fronts, not just in individual, spectacular moments of triumph. In the climate battle, the frontlines are everywhere, which magnifies our choices and gives us many directions to choose from, even if it can make impacts less visible. Sometimes, the work we do will be paid, but sometimes it will not. It may take years to polish off the layers of other people's expectations of us that have covered up the true form of our calling. Learning to focus on what you love more than on what you think you are *supposed* to pursue will help you figure out where to devote a lifetime's worth of energy.

Tease out your relationship to the work that needs to be done. There is no one-size-fits-all prescription. On the contrary, the task is to do the self-study required to figure out what satisfies your emotional needs and suits your limits and strengths. Any of your own individual responses, no matter how small, "has its own power to shift the balance" in some sphere of influence, says educator Marie Eaton.

We *Are* Too Small, but Small Is All

You may have grand dreams. They might include an overhaul of capitalism, a 50 percent reduction of the world's population, universal government subsidies for green technology, or the granting of legal rights to all species. But the scale and scope of these tasks may make you feel small, weak, and defeated, even before you've started trying. These sensations may be exacerbated by the "imposter syndrome"—the feeling that you're not qualified to do anything meaningful in the first place; young people notoriously feel they have no power, a condition that has been shown to be more intense for females and people of color.

But to be a change agent, you don't need spectacular results. When you're feeling demoralized, remind yourself of adrienne maree brown's tenet of emergent strategy: "Small is good, small is all." We may feel individually too small, but as brown goes on to say, "Small actions and connections create complex systems." And gather inspiration from the famous statement attributed to anthropologist Margaret Mead: "Never doubt that a small group of thoughtful, committed citizens can change the world. Indeed, it is the only that thing ever has."

My student Ezzie began with small steps. She approached her congregation about making her church's garden more accessible to her Latinx community. This strengthened her bonds to her church and to her rural migrant community. By making the garden welcoming to Latinx members of the church, not just the white members, Ezzie helped empower her church and community members to take food production into their own hands and express their cultural traditions more freely. It was a revelation for her to realize that she had become a social change agent in her community.

Biographies of social change agents such as Martin Luther King Jr. can give us the impression that only those on impossibly high pedestals can achieve anything worthwhile. Watching the news, we are further duped into thinking that social change is driven by the people who speak the loudest, do the most grandstanding, and bully others into doing their bidding. In what activist and writer Paul Rogat Loeb describes as "collective amnesia," we forget about the ordinary people—from conservationists to suffragists to abolitionists to civil rights activists—who participated in grassroots movements and were able to shift public sentiment and challenge structures of power.

If we think that social change happens only by individual feats in triumphant bursts, if we think our power lies in being consumers rather than social change actors, and if we refuse to learn from social movement elders about how they cultivated resilience against all odds, then our collective amnesia will only amplify the scariness of climate change.

The remedy is finding a sense of purpose by gaining the intellectual and emotional confidence to plug in and build relationships with others who are doing the same. This will take cultivating confidence in ourselves as agents of social change.

Find Your Sphere of Influence

Let's say you think the best solutions to environmental ills are to elect new politicians, transform capitalism, or limit your waste stream—changes in one or more of the political, economic, or consumer realms. These structures are inextricably linked with other structures, which opens us up to imagine a wider set of points of intervention. Economist Donella Meadows calls these "leverage points": "places within a complex system (a corporation, an economy, a living body, a city, an ecosystem) where a small shift in one thing can produce big changes in everything." Greta Thunberg's Friday school strikes, beginning in August 2018, are illustrative. By no measure could we count one young girl skipping school to protest alone in front of the Swedish parliament as a large event, but the small act snowballed into a global movement, which had already been congealing in myriad places and by great effort by lots of people around the world. I see students working on all of these seemingly small efforts, pushing leverage points every day, sometimes by collectively planning and organizing—to run a workshop for a community group on privilege in the environmental movement, for example—and sometimes by putting their solitary voices out into the world, as when a student creates a blog for a school assignment and, from the online responses, realizes she's not the only person who thinks and feels the way she does. The "small" successes reinforce confidence and build students' sense of being in a community of others. When this happens, worlds start to shift, and sometimes cataclysmically, as we are seeing now.

Leverage points are everywhere, and we already have networks, relationships, and influence that we can put to good use in our climate work. Start by mapping out your "spheres of influence," as psychologists and organizers call them (figure 4). When you begin to see all the people with whom you have reciprocal relationships, and all the potential points of intervention in the systems in which you function daily, you can recognize the ways that you already have a significant amount of power.

Changing hearts and minds takes work. Changing structures takes work. Changing culture takes work. Changing policy—even the policies that we are directly in contact with in our schools, families, churches, and other communities—takes work. On all of these fronts, we are already doing work, and more work can be done.

Resilience is key to staying the course and remaining effective agents of change over the long term. In *Transformational Resilience,* social systems theorist Bob Doppelt argues that there are three core elements to approaching climate disruption: reducing emissions (also known in the climate movement as mitigation), preparing infrastructure to be resilient to climate change (adaptation), and

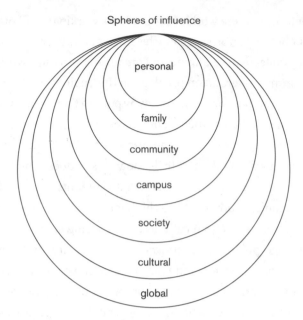

Spheres of influence

personal

family

community

campus

society

cultural

global

FIGURE 4. Spheres of influence. Reprinted by permission from Abigail Reyes.

"building the capacity of people to cope with the traumas and chronic toxic stresses generated by rising temperatures and use them as catalysts to learn, grow, and increase personal and collective wellbeing, and the condition of the natural environment." In "Rethink Activism in the Face of Catastrophic Biological Collapse," journalist Dahr Jamail and organizational change consultant Barbara Cecil similarly posit three domains of activism in this unprecedented historical moment: fixing the mess, mitigating inevitable suffering and loss, and entering the realm of "healing, reparation, . . . and collaboration." While all of these domains are necessary, we do not have enough models or ideas about how to achieve the third element in these scenarios—personal resilience—which is what I'm trying to address in this book. Based on my own experiences, I agree with Doppelt: "No response to the climate crisis will . . . succeed unless individuals and groups of all types around the globe understand how trauma and toxic stress affects their minds and bodies, and use skills to calm their emotions and thoughts, learn from, and find meaning, direction, and hope in adversity."

Our sense of efficacy in the face of seemingly insurmountable obstacles rests on redefining what we think of as meaningful actions in the world, and what we think of as our role in that work. Let's start with shifting our perception of what an "impact" is, how big it needs to be, and whether any one of us can make enough of it. "First, we need to believe that our individual involvement is worthwhile," Loeb reminds us, and "this is as much psychological as political."

The Myth of Measuring

Social movement theorists, activists, and leaders have long puzzled over the dilemma of whether it is worth doing anything in support of a cause when it is not clear whether that effort will achieve the outcome desired. Why sacrifice so much when there's no guarantee that the vision will be realized? The journalist I. F. Stone had an illuminating response: when asked why he kept writing critically about US involvement in Vietnam even though no one was listening to him, he said, "If you expect to see the final results of your work, you simply have not asked a big enough question."

Teachers know that attempting to measure the effects of our actions is a pointless exercise. Although we are required to prove our effectiveness, it is clear that our impacts go well beyond what we can gauge with exit surveys and accreditation assessments, and that all of these impacts may percolate over long periods of time, outside any reasonable frame of measurement. Author Rebecca Solnit warns us of this fetish to measure impact when she describes, for example, how the events of World War I led, fifty years later, to the 1960s nuclear accord. She sees social change happening in small, often unglamorous, even invisible stages: "It's always too soon to go home. And it's always too soon to calculate effect." We often expect that "for every action there is an equal and opposite and *punctual* reaction, and regard the lack of one as failure." But change doesn't work like that. If we need evidence of impact, we will burn out waiting for our utopia to manifest. "The world gets better. It also gets worse. The time it will take you to address this is exactly equal to your lifetime, and if you're lucky you don't know how long that is. The future is dark. Like night. There are probabilities and likelihoods, but there are no guarantees."

Similarly, author and activist Howard Zinn reminds us that "revolutionary change does not come as one cataclysmic moment (beware of such moments!) but as an endless succession of surprises, moving zigzag toward a more decent society. We don't have to engage in grand, heroic actions to participate in the process of change."

These seasoned activists' ideas about social change can relieve us of "big success" expectations, which cause existential suffering. Social change is slow, un-newsworthy, and unspectacular. Being an agent of social change therefore requires humility and offers an opportunity to explore how much our egos motivate our efforts.

Evidence of success has become part of our cultural ethos. In the instrumentalist view, the only efforts worth making are ones that yield immediate, obvious results. While deliverables and metrics produced via evidence-based strategies may be worthy in organizational culture, they probably should not be applied to a life's work. Social movement leaders from Gandhi to Thich Nhat Hanh know that instrumentalism is a red herring. Social change is hard to measure. Triumphant moments are the result of the invisible work of scores of unnamed idealists and of forces impossible to track.

This is grounds for hope, not despair. The work, not utopia, is the endgame. Loeb describes the thought process of activists who hold themselves to impossibly perfect standards: "Even when we know better, we sometimes feel we have to tackle everything at once. If our efforts don't instantly achieve dramatic results, we are quick to criticize ourselves, and doubt that our efforts can matter." As a solution, he suggests we aspire to being "good-enough activists" who don't demand perfection or certainty before taking a stand, "realizing that although we may never win the Nobel Peace Prize, our contribution can still make a difference."

Literary theorist Nicole Seymour argues that environmentalists would do well to recognize the irony that underlies their entire mandate. *Irony*, I believe, is a much more fun orientation to our work than *problem-solving our way out of despair*, which is the dominant—and, in my view, dreary and cynical—environmentalist mode. Social movements that span generations, such as the climate movement, have to have a long view of social change.

Redefine Action

The myopic focus on action at the expense of theorizing, cultivating collective and personal resilience, and even plain-old contemplation is also a kind of anti-intellectualism. It is part of a broader cultural shift away from valuing the world of ideas and thinking about the existential questions of life, and toward making intellectual activities seem frivolous or only of interest to the privileged. In the worlds of environmentalism and environmental studies, I notice a strong anti-intellectual tendency. Many people want to be problem-solvers and to fix things right NOW. They want less feeling, less thinking, less talking, and more action. The urgency of the situation seems to demand it. But urgency and its sidekick, non-thinking, result in unintended consequences that can undermine our goals.

Actions that emerge from impulse or the "fight or flight" parts of our brain are likely to exacerbate problems, for example, by making it hard to see structural causes and making us want to blame the wrong people. The rapid rise in global refugees and the crisis at the US-Mexico border are powerful examples. It might be tempting to say "fix the problems in your own country and stop coming to ours," but this approach ignores the larger causes that make people move, such as histories of colonialism and neoliberal trade policies, which aggravate resource scarcity and political instability, or rising sea levels caused by global emissions and climate change, caused mostly by rich countries. Building walls is a knee-jerk reaction; in contrast, thinking through these causes and effects can help us identify structural solutions. We waste a lot of energy and resources throwing up defenses in the wrong places and reacting impulsively to every threat that crosses our bows. Coercive conservation, which I discuss in the next section, is an example of how well-intended actions can go wrong when their consequences aren't thought through.

Instrumentalism—the belief that ideas are important only insofar as they are useful— also makes it easy to overlook less obvious forms of social change. What counts as a meaningful action is entirely contextual. In a certain situation, action taken might not seem to be enough—but it may be all that is possible. If we follow our narrow definitions of action through to their logical conclusion of meaningful accomplishment, it would seem that there is never an end; it will never be enough. We can fill a black hole with a frenetic outpouring of effort, but that is not a prescription for lifelong resilience.

Yet there is action in things that we are not accustomed to thinking of as arenas of social change. Is changing somebody else's mind an "action"? Is changing your own mind an action? Is *inaction* ever an important form of action? What assumptions shape our ideas of what counts as meaningful action, and do we want to perpetuate them? For example, adrienne maree brown suggests that the need for dramatic, visible action is a symptom of patriarchy. To fetishize action and reject the less visible and less glamorous work of caregiving, community organizing, or going to meetings is to subscribe to a gendered view of the kind of work that matters.

If urgency is driving your emphasis on action over other priorities, then you might consider rethinking its value. Buddhists understand the interdependence of thought, feeling, and action. The Eightfold Path is about getting the feelings and the thoughts "right" for "right action." If what counts as action for climate justice for you is changing a law or dismantling capitalism, then you may never feel adequate, unless you can break these goals down into next steps that are much smaller and will require exactly the opposite of urgency—patience. Bigger things don't happen without many other smaller and simultaneous changes. Bringing about change is a daily leap of faith.

Seymour lists many actions that the instrumentalist approach to environmentalism overlooks or does not explicitly value:

expressing dissatisfaction or *disaffectation* with the environmentalist status quo,
bearing witness to crisis,

enacting catharsis,

raising activist morale,

building community,

serving as cultural diagnoses,

indexing and helping us understand our current eco-political
moment, . . .

mitigating the "partisan divide over environmental matters,"

inculcating a new range of responses to crisis, . . .

modeling flexibility and creativity in the face of crisis, and

inspiring . . . "artful endurance."

All of these non-instrumental actions are worth doing, and anybody can do them, right now. They broaden our notion of what counts and make it apparent that we indeed do have capacity to act effectively. Anybody who rejects "artful endurance" or "bearing witness" as too passive, minor, or frivolous is missing an opportunity and surrendering their imagination—and their power.

A perception of ourselves as powerless emerges not from our actual inefficacy, but from our ideas of what it means to have the capacity to shape social change. This view is caused by the instrumentalist conceit that the only actions that matter are the ones that make impressive, immediate, large-scale change. The problem is not that we have no power. Rather, the problem is that we don't use the power we *do* have.

Our sense of powerlessness also comes from the way climate change is framed: as "distant in time and space, invisible to the senses, abstract, and socially far from our influence," in the words of Per Espen Stoknes. The delayed effects of climate change will continue for a very long time, and mitigating them seems beyond our "locus of control." But you can scale out the things you do within your spheres of influence. You may not be able to make every country in the world, much less your own, abide by carbon taxes, but you can build incentives into your own household to pollute less, such as reducing your waste stream. Or you can preserve resources by practicing water catchment at home. These efforts could help you get an internship or a job at a nonprofit that educates people on food waste or promotes household water catchment systems, plugging you into a passionate and supportive network of people who geek out on saving the planet in the same ways you do. By focusing on our skills and joining networks, we can see our actions ripple out.

When we think of action, we assume its lesser opposites are thought and emotion. The emphasis on action prioritizes the "hand" over the roles of "head" and "heart." But these other two are as important for climate justice as the first. To balance out the bias for action, approach the actions you take with emotional intelligence and strategy.

If we enlarge on our idea of what counts as transformational action, we might imagine ourselves as not only nudging the movement forward, but even providing leadership. In *ReGeneration: Young People Shaping Environmental Justice,* authors Julie Quiroz-Martinez, Diana Pei Wu, and Kristen Zimmerman offer conclusions from the Second National People of Color Environmental Leadership Summit (2002), which redefined leadership as founded in facilitation, community-building, and bottom-up organizing, not top-down fixes.

In their view, a leader does the following:

- Facilitates
- Acts as a role model
- Is accountable to the community

- Listens to the people
- Is open minded
- Takes initiative
- Contributes and participates
- Represents the voices of the people
- Serves the people
- Is culturally sensitive to the differences of people and doesn't push her/his ways onto others
- Is organized and wants to organize
- Takes responsibility for his/her actions
- Is honest
- Practices what s/he preaches

We can all incorporate these leadership principles into our day-to-day lives.

Or we may elect to be gadflies. One of my dearest scholar-activist colleagues said to me once, "I want my students to be problem-makers, not problem-solvers." She encourages them to challenge how issues have been defined by mainstream media, and even environmentalists, and to use cognitive tools to examine how problems are intertwined before acting on them. Such "systems thinking" can connect the dots between our efforts to address climate justice and, for example, the prison industrial complex, which relies on both ecological and human disposability.

Slow Hope

Urgency driven by fear can make us do impulsive, irrational things. As an antidote, environmental philosopher Timothy Morton wrote an essay titled "Don't Just Do Something, Sit There!" In it, Morton argues against the binary of action/thought, and offers "ecological thought" as a *kind of action:* "Reframing our world, our problems, and ourselves is part of the ecological project. This is what praxis means—action that is thoughtful and thought that is active." Ecological thought—awareness of connection—can "call us from within the grief." Morton's concept of ecological thought emphasizes contemplation over simply diving in, which can easily make things worse.

So ignore Nike, and don't "Just do it." Instead, heed environmental literary scholar Scott Slovic, whose book *Going Away to Think: Engagement, Retreat, and Ecocritical Responsibility* stresses the importance of damping our heroic problem-solving impulses in favor of deliberation and thoughtfulness. "We never quite fathom the true magnitude of our influence on the world," he writes, but the value of "slow thinking" and "moral reflection suggests that the potential for positive, life-sustaining impact is in itself motivation to act." Christof Mauch writes about the need for "stories of slow hope" to remind us that change is incremental and not always obvious.

Such stories rarely make the news, however. The painstaking effort that was required to protect a particular eelgrass habitat in Humboldt Bay, a nursery for biodiversity and a climate change–mitigating carbon sink, is an example of "slow hope." It took a long time to create, was unglamorous work, involved few identifiable heroes or obvious triumphs, and is otherwise hard to "see," because the eelgrass is just still there. Or take the "international outpouring of marine conservation success stories" collected by the Twitter initiative #OceanOptimism, which showcases how people are creatively removing plastics from the oceans, organizing to protect marine areas, and modeling conservation collaborations. Slow down, find stories of slow change, and cultivate slow hope.

Urgency and action without mindful deliberation does not serve our mission. As the racial justice organization Dismantling Racism puts it, a continued sense of urgency makes it "difficult to take time to be inclusive, encourage democratic and/or thoughtful decision-making, to think long-term, and to consider consequences." It frequently involves the sacrificing of potential allies in favor of faster results, and undermines the interests of communities of color "in order to win victories for white people (seen as [the] default or norm community)." In the case of climate action, the urgency mode of the problem-solving ethos constrains the movement's ability to attend to social justice and to be deliberative about negotiating courses of action. Advocates of climate justice, for example, caution against pushing for green jobs without taking time to transition workers in fossil fuel industries to new jobs. In our pursuit of evidence-based impacts, driven by the urgency of the planet's demise, we also limit our potential to create coalitions in new places, and to solve all kinds of other problems we don't even know exist.

"Coercive conservation" describes how, in the name of natural resource conservation, communities across the planet have been booted out of their homelands. In the late nineteenth century, conservationists in the United States removed indigenous peoples from areas they wanted to turn into national parks. The "urgency" was that the so-called American frontier had run out, and there was no more open land to act as a safety valve for the social and cultural growing pains of a new, urbanizing, industrializing, diversifying country. In the name of securing permanent spaces for (white) Americans to maintain the unique American concept that open space makes good citizens, the National Park Service was born, at great cost and violence to the people who had lived in those spaces for millennia. Support for the value of untouched nature by such notables as John Muir made it easy for audiences to demonize those who lived in or used that nature, and to enforce by any means necessary the conditions for cultivating a distinct American national identity.

Urgency and action come from admirable places in us—a desire to help and an awareness of the direness of our situation. But we must be wary of putting all our eggs in the action basket. Reflection, emotion, and deliberateness are good for social justice and effective action-planning—and they will keep us from burning us out existentially.

The Ocean in a Drop

In *Blessed Unrest,* activist and journalist Paul Hawken estimates that as of 2007 there were at least one and possibly over two million organizations worldwide working on issues of environmental and social justice. He calls this the largest movement in all of human history. We are not alone in our efforts.

If we see ourselves working collectively rather than individually, we can rest assured that we are contributing to a larger web of movement, what adrienne maree brown calls "emergent strategy," spiraling out from even the smallest of our actions— a "relational, adaptive, fractal, interdependent, decentralized, transformative" way to facilitate, direct, and understand social change. Envisioning a compelling future is central to emergent strategy. The conviction that our efforts are multiplied by a larger community that we may have never met emboldens us to keep doing the work and staves off the discouragement we may feel when we look around and see so much destructive behavior. Hawken assures us that the movement is spreading in every country and through every walk of life, and asks us to imagine that we are in the company of English bird-watchers and French farmers, indigenous tribes in Bolivia and housewives in Japan, as the movement "is taking shape in schoolrooms, farms, jungles, villages, companies, deserts, fisheries, slums—and yes, even fancy New York hotels."

Knowing that we are part of a collective gives us permission to rest. We can recover from our exertions knowing that others, who also have taken care to sustain themselves, can take over the work. We all must take care of ourselves so that we can step up when others need to tend to themselves. The perception that social change happens only on an individual scale creates defeatism. *Of course* we cannot solve the problems by ourselves. It's worth checking in occasionally on all the engagement and talent out there—what some call asset or capacity mapping—and to gather stories of people succeeding—to surround ourselves with lists of collective achievements. As brown writes, we must "let that work be tangible in a way that shows how the work builds over time."

Leslie Davenport adds, from the perspective of a psychologist, that "emphasizing the power of collective action builds confidence that change is possible." Instead of focusing on news sources that show us the terrible sides of humanity, which makes us feel isolated and suspicious of others, seek out stories about how social change is occurring, learn about how movement leaders cultivate community, and treat the tending of relationships as an essential, serious form of social action, not just a personal need. A new phenomenon, "solutions journalism," has arisen to counterbalance the dominant media's emphasis on intractable problems with stories of people actively solving problems; the journal *Solutions* and the trade magazine *Yes!* are two that make the work visible. Examples of positive change are elusive not because positive change doesn't exist, but because it does not capture our attention and fails to meet media criteria for news.

There's evidence that our collective efforts can make a greater contribution than material resources to a community's ability to withstand climate change. According to Davenport, "it is social capital, our connections to and caring for each other—not the money poured into infrastructure—that most quickly leads to recovery following a climate disaster." Bob Doppelt similarly points to the "growing body of evidence" that "building personal resilience skills, robust social support networks, and close collaborations among local organizations" are the "keys to enhancing human resilience," more than "physical infrastructure, natural resource adaptation, or post-trauma treatment."

If a dollar spent on building social capital will go farther than a dollar spent on infrastructure, shouldn't we be investing in community resilience and trust over the long haul instead of applying Band-Aids in moments of crisis? We need social capital as much as or more than we need dikes and levies. It feels much more doable to contribute to robust social support networks than to get everyone on the planet to limit their emissions or hold back the rising seas. If we shift our notion of what the solutions are—not simply infrastructure but also interpersonal trust—we can focus on the latter as we mobilize collectively to demand the former.

The myth of rugged individualism undermines resilience by making us think we are alone in our efforts as well as in our feelings of despair. It tells us that our actions, be they good or bad, don't spread beyond our immediate purview, that we are absolved of the upstream and downstream consequences of our behavior, and that we have no power to effect change. brown turns this around, urging us to "focus on critical connection more than critical mass—build the resilience by building the relationships." Measuring success by the strength of our relationships is far more likely to result in fulfillment and a salve for our existential isolation.

Cognitive psychologists recognize the importance of being aware that we are part of a team. "People easily feel helpless if left on their own when confronted with the severity of the coming climate disruptions," Stoknes observes. But "participating in a community or group that works for a common cause is a good remedy (the only one, actually) for this toxic helplessness and passivity." When my students recognize that they are all in the same boat and that they *need* to rely on each other, to express vulnerability about their anxiety and dread, and to cultivate community morale, they become the change they want to manifest in the world. They spread that uplift to their other classes. Not feeling alone is probably the most important prescription for long-term resilience. When our classes start with building community before learning content, we all have far more energy and passion about the subject matter when we finally get there.

A variety of other kinds of expertise and ways of thinking also emphasize relationality over individualism. Indigenous scholars such as Robin Wall Kimmerer articulate how both scientific and traditional knowledge recognize the mutual dependence of living matter, reflecting a "reciprocity" of actions between the environment and ourselves. Forms of environmental theory like "new materialism," as described by Jane Bennett and Stacy Alaimo, and "eco-phenomenology," as described by David Abrams and Ted

Toadvine, insist on our interrelationship with nature, right down to the cellular level. Bennett draws on science to suggest that everything, including things we've been taught are inanimate, like rocks and desks, are in fact "vibrant matter." Buddhists such as Thich Nhat Hanh espouse a notion of "interbeing," which can be a foundation for acting in right relation with the world. Even evolutionary biologists like Lynn Margulis argue that our very bodies are the result of symbiotic relationships among various forms of life. And intersectional activists articulate this insight in terms of the struggle for self-determination. As indigenous activist Lilla Watson said, summarizing the sentiment of Aboriginal activist groups: "If you have come here to help me, you are wasting your time. But if you have come because your liberation is bound up with mine, then let us work together."

Seeing the world as interdependent is consistent with Morton's "ecological thought": relation isn't a choice, it's a reality. Why choose to see yourself as one individual facing a world of evil? This perspective is counterproductive both for you and for those around you. Rather than dwell on the ways that climate deniers are different from you, rousing feelings of disgust (see chapter 5), alarm, and righteousness, why not focus your mind on all the vibrant matter—human and more-than-human—with which we are in reciprocal relation? The world you love and rely on is right in front of you; protect it by tending it, not by raging against the things that threaten it.

Powerlessness Is an Illusion

The concept of "pseudoinefficacy" illustrates how we come to think of ourselves as powerless. Pseudoinefficacy is a way of thinking that we can have little positive impact on the world, which in turn shapes our decisions about where we will devote our energies. In a thought experiment described by philosopher Peter Singer in "The Drowning Child and the Expanding Circle," Singer asks his students if they would feel an obligation to save a child who was drowning in a pond they are walking by. They unanimously say they would. He then widens the circle, asking if they would feel an obligation to save a child in another country who was at equal risk of death, at which point they become more hesitant; from there, he goes on to raise questions about our moral responsibilities for the lives of others. Cognitive and conservation psychologists Daniel Västfjäll, Paul Slovic, and Marcus Mayorga pose a new question: "Suppose, as you see the child go under, you also see, in the distance, another child beginning to drown—one you cannot reach. Would you then be less motivated to rescue the child within your reach?" Their research suggests that because it would not be possible to help both, people are less likely to attempt to save even the first child. This makes no rational sense. Isn't the incentive—the benefit of a life saved—the same in both cases?

Västfjäll and colleagues use the poet Zbigniew Herbert's phrase "arithmetic of compassion" to describe how the positive feelings of saving a child are canceled by the negative feelings of knowing that others cannot be saved. In other words, the desire to help fades when we realize we can't help everyone who needs it. For those in a position to offer assistance, "decisions are strongly motivated by perceived efficacy. Inefficacy, real or perceived, shrivels compassion and response." Feeling like you don't have power to do good will deflate your desire to even try. It follows that one's efficacy, or confidence in one's ability to solve a problem, is more likely to determine whether one even attempts to solve a problem.

More than having marketable skills or being armed with information and arguments, *believing in your efficacy* will influence whether you try to fix the problems you see. This is the premise of this book, and I cannot overstate how key this concept is for all of the arguments I make here. It will affect how much stress and depression you experience in threatening situations, which in turn affects your motivation to do anything about that situation. "Perceived rather than actual efficacy is often the determinant of behavior," Västfjäll, Slovic, and Mayorga argue. "The problem of pseudoinefficacy," they continue, "is central to a wide range of important personal and societal decisions motivated by perceived efficacy, such as actions to mitigate climate change or other threats to human health and the environment." In other words, thinking of ourselves as ineffectual in our ability to address climate change *makes us so.*

In turn, if we feel that our actions will never be big or impactful enough to solve the problem, we are adopting a drop-in-the-bucket imaginary that causes feelings of despair and hopelessness. This insight is hugely important for climate justice advocates. The negative emotions that come with knowing that we cannot fix climate change will reduce the positive emotions we might induce by doing some smaller action. The very *scale* of the problem makes people not want to do anything about it. Considering climate stories through this lens, it's astonishing that so many climate change advocates still rely on the sky-is-falling approach to getting people to care about climate change. Perhaps they think that heralding the end will propel more listeners to action, and thereby help us avoid that end. But cognitively, it doesn't work this way, or at least, it won't in the long term. The bigger the problem, the less fixable it seems, and so the more likely we are to do nothing instead of something. We are finished before we even start. To fix this, we need to actively combat messages that tell us that the problem is too big to fix, and to remind ourselves that small is all, and that small is enough.

Manifest(o)ing Your Yes

Social change work can be tedious, tiring, and hit or miss. When people who have been doing it for a long time come to talk to my students about how they rebounded from hopelessness, failure, and exhaustion, my students feel energized to dive back into their work, too. Knowing that this is how social change works helps us accept its challenges and find some pleasure in the messiness of it. When Ron Finley came to campus and explained that in South Central Los Angeles, the land of "drive-throughs and drive-bys" (suggestive of the parallels between the violence of fast food and the violence of guns), providing access to fresh food was both a matter of racial justice and environmental health, his story helped my students see how activists have long practiced forms of resilience in order to make better worlds.

The task of combatting climate change seems huge. Your to-do list may already be so overwhelming that your pulse races every time you imagine one more thing to do to save the planet. A strong sense of purpose helps us filter out the tasks we can say no to. Believing that every demand for our attention is worth heeding is a recipe for burnout. Every committee needs more members, every protest needs more bodies, every person we know needs support, every representative needs to be called every day, every turn of events needs your social media exposure, every Facebook thread needs a comment, every email a response. The vast number of arenas in which we could show up and act is stupefying, and may feel even more so when we accept the principle that the frontlines are everywhere and that you can make differences in all kinds of places. The need for our energy and time is a bottomless pit. What we do have control over is our ability to say no, so that when we say yes, we really know that the work we do is aligned with our mission and will reinvigorate us.

brown advises, "get really good at being intentional with where you put your energy, letting go as quickly as you can of things that aren't part of your visionary life's work. Then you can give your all, from a well-resourced place, when the storm comes, or for those last crucial miles." The self-study required to get good at being intentional with our energy takes time and is not a linear process. It may start with brushing your teeth with your nondominant hand. As we change and evolve, the object of study—our self— becomes a moving target. The first step is simply to take our emotional responses to climate change seriously instead of thinking we "should" be activists or politicians, or that we "should" do more.

I make a practice of asking my students to write their own "manifestos," or mission statements, some of which are published on a blog called "Critical Hope" that I created for seniors as they prepare to graduate. Everybody should write a manifesto, and repeat as needed as your life advances. I invite you to write one now. Take ten minutes to free-write about why you care about the planet, why you care about suffering, and what you personally are uniquely positioned by skill or passion to do in the world. Your declaration of intention and desire will be a beacon to return to when you feel lost or overwhelmed. The simple act of writing your eco-mission statement can move you in the direction of manifesting it.

Once written, you can ask of every decision you make, "Does this choice serve my mission?" Rather than getting distracted by social media, procrastinating, or sabotaging yourself, you can make daily decisions that will cumulatively add up to creating the future you desire.

You don't need to love animals or even like being in "nature" to make a difference in addressing climate justice. It's not a prerequisite to helping save the planet that you be good at math or love gardening. You don't have to be more powerful than you already are. You can start simply by identifying and using the powers that you already have.

Whether we are "artivists" or trauma therapists or students or stay-at-home-dads, whether we work on reducing emissions (the science side) or on voter suppression, gerrymandering, or toxic siting (the social justice side), we have the capacity to address the social structures that create environmental injustice and aggravate climate change. The frontlines are the books we choose to read to our children, the pipeline being laid across our land, the social capital we foster among our coworkers at McDonald's, the cli-fi novel we have been fantasizing about writing, the campaign to elect a pollution-fighting candidate, meditating, or listening compassionately to somebody with whom we disagree.

Climate justice needs all kinds of help. Scientists are only one part of the groundswell, just as we each are only one part. If the problem of climate change is wickedly complicated, we will need people everywhere and with every type of skill. Following brown, know that "uprisings and resistance and mass movement require a tolerance of messiness, a tolerance of many, many paths being walked on at once." What path will you walk?

Finding Your Climate Action "Thing" and Doing it Well

☐ Examine your spheres of influence, sites of change, and leverage points.

☐ Ecosystems thrive on diversity and symbiosis; so does the movement to preserve them.

☐ Cultivate a radical imagination.

☐ Resist measuring.

☐ Accept messiness.

☐ Beware of the cognitive trickery of perceived inefficacy.

☐ Slow down, find positive stories of slow change, and cultivate slow hope.

☐ Know what to say no to, so you can be fully present for what you say yes to.

☐ Value quality over quantity in your social investments; critical connection > critical mass.

4   Hack the Story

"More Americans can imagine the 'end of the world,'" says sociologist Kari Norgaard, "than can envision a switch from using fossil fuels or an economic order other than capitalism." If it's easier to imagine the end of the world than it is to visualize the transformation of existing economic, cultural, and political structures to combat climate change, then doomsday may become a self-fulfilling prophecy.

The dramatic imagery of apocalypse that is transmitted by news media and social media exaggerates the negative aspects of the future. Headlines blare "We're Losing the War on Climate Change" (CNN), "End of Civilization: Climate Change Apocalypse Could Start by 2050 If We Don't Act" (*USA Today*), "The Insect Apocalypse Is Here" (*New York Times*), "Climate Change and the New Age of Extinction" (*New Yorker*), "Time to Panic" (*New York Times*).

Television, radio, newspapers, and the internet feed us bad news in part because we seek it. "If it bleeds, it leads," as the saying goes. Although people say they generally prefer to hear positive news, scientists have found that when given a choice, people will more often select—and remember—depressing stories. As I explain in this chapter, this is chiefly because of the way humans have evolved to respond to threats. Our minds intuitively seek negativity, news outlets leverage this susceptibility, and as a result, unless we're holed up in a cave, we act and live from an orientation of fear.

When our worst nightmares seem to be coming true, we may want to hide our heads in the sand and abdicate our responsibility and power. The apocalyptic narrative makes us look for villains and heroes, ignore gray areas, and reject imperfect but meaningful proposals for progress.

Environmental scientist and educator Elin Kelsey points out that the lack of stories to help us imagine a different future makes the apocalypse all the more likely. "Environmental solutions are emerging and maturing all over the world. But they are often ignored or trivialized or underreported and thus the likelihood of replicating them is reduced because they do not fit the battle cry we constantly hear—the Earth is doomed." The environmental movement itself, because of its determination to goad people into action, has been guilty of peddling pessimistic and discouraging stories.

What can we do to counter the effect of the disaster scenarios that run on endless replay in our minds? In this chapter, I describe how we can develop habits of attention that serve us better, learn healthy ways to privilege facts over feelings, distinguish between positive and negative narratives, and create our own stories to guide and sustain ourselves.

Ursula K. Le Guin speaks eloquently about our storytelling potential: "We live in capitalism. Its power seems inescapable. So did the divine right of kings. Any human power can be resisted and changed by human beings. Resistance and change often begin in art, and very often in our art—the art of words." We will never obtain the social and economic structure we want if we don't vigilantly cultivate the ability to imagine it.

Our Psychological Makeup

The news media leverages our innate propensity to zero in on negative information. As cognitive psychologist Steven Pinker writes, "Whether or not the world really is getting worse, the nature of news will interact with the nature of cognition to make us think that it is."

Environmental problems receive public and political attention to the extent that they are perceived as a risk. But our perceptions often don't match the probability of actual threats occurring. Cognitive psychologists have shown that humans pay the most attention to immediate threats that involve frightening, potentially malevolent beings such as a bear lurking behind the shrubbery or somebody poised to break into your house. We also perceive threats as greater risks if they have a face and are perceptible or visceral. A burglar is scarier than a tree falling on our house, but a tree falling on our house is more visceral than mold growing in the basement. And we're less likely to be afraid of driving (because we feel in control behind the wheel) than flying (because we surrender control to a pilot), even though driving is statistically more dangerous.

Climate change, in contrast, is not a close encounter with a malevolent being, and its negative effects will be perceptible mostly over time. The fires in California, the hurricanes in the Gulf—these are the scary faces of climate change; and yet the more slowly or subtly developing changes—the destruction of marine ecosystems, melting icecaps and glaciers, changes in atmospheric flows, human displacement, disruptions to our food supply—may have greater consequences. We struggle to translate these protracted developments into an immediate threat to life and limb. Anthony Leiserowitz, director of the Yale Program on Climate Change Communication and an expert on risk perception, has said about climate change, "You almost couldn't design a problem that is a worse fit for our underlying psychology or our institutions of decision making."

Social media fuels the perception that these threats are likely to come to pass. When we see waves of shootings, wildfires, drought, refugees, and insect-borne disease on our feeds, we're more likely to be afraid of those things happening. This is largely due to the "availability heuristic," a phenomenon uncovered by psychologists Amos Tversky and Daniel Kahneman that leads people to "estimate the probability of an event or the frequency of a kind of thing by the ease with which instances come to mind." Our fear isn't based on reason or statistical evidence, but rather on this cognitive mind trick. If we've seen a representation of it, we can imagine it, so we believe it to be true.

Thoughtful, reasoned analyses of climate change rarely make the news. In *Slow Violence,* environmental literary critic Rob Nixon argues that the media's focus on sensational, spectacle-driven messaging means that hurricanes make the news, but not the history of infrastructure and policies that make hurricanes a disaster, nor the untold years of recovery afterward. Positive events are also less likely to be covered, since good news attracts fewer viewers and doesn't have the shock value that the media covets. Steven Pinker puts it another way: "Bad things can happen quickly, but good things aren't built in a day, and as they unfold, they will be out of sync with the news cycle."

We are more afraid and cynical than it makes sense to be. There is a lot of good news to celebrate, yet news outlets around the world are gloomier than ever. Pinker warns that "the consequences of negative news are themselves negative. Far from being better informed, heavy newswatchers can become miscalibrated," which makes them worry more. Pinker's solution is to develop a quantitative mindset to see more clearly what is really happening in the world.

But then we have to deal with the fact that even when people are presented with actual data, they are inclined not to believe it.

Is climate change really as grave a situation as it's made out to be? Is it more serious than Vietnam, for my own parents' generation, or the threat of nuclear war before that, or the Holocaust before that? Yes and no. On the one hand, consensus is building that climate change is the worst crisis humanity has ever faced. On the other hand, other aspects of our lives—including health, safety, and income—are on the upswing. Taken together, it would seem that climate change is as bad as scientists report, but it's happening against a backdrop of overall improvement of the quality of human life.

In *Factfulness: Ten Reasons We're Wrong about the World—and Why Things Are Better Than You Think,* Swedish physician Hans Rosling combs through long-term data measuring well-being, from the proportion of people who are enslaved to the number of girls who are receiving an education. His conclusion is that the state of the world is improving. However, he also reports the fascinating truth that *the vast majority of educated people underestimate the improvements.* Rosling deduces that our brains are hardwired with instincts that cause us to misperceive facts. He has identified ten instincts that tend to distort our perceptions of the world:

- Gap instinct: We divide things into two discrete groups with an imagined gap between them (whereas most things usually fall on a spectrum).
- Negativity instinct: We notice the bad more than the good.

- Straight-line instinct: We believe that progress is linear—for example, that a line on a chart will continue to go in the same direction.
- Fear instinct: We focus on information that triggers fear.
- Size instinct: We see things out of proportion and exaggerate the importance of a single piece of data.
- Generalization instinct: We group things and people in categories.
- Destiny instinct: We believe that innate characteristics determine the destinies of people, countries, religions, and cultures.
- Single-perspective instinct: We are attracted to simple causes or solutions.
- Blame instinct: We want to find a scapegoat when bad things happen.
- Urgency instinct: We believe we have to act NOW.

Rosling suggests opposing our "overdramatic worldview" with a "factfulness worldview" to help us override the cognitive instincts that keep us locked in dread about the future. The key here is the application of critical thinking. Rosling supplies a cheat sheet to help us realistically examine the stories we encounter:

- Look for the majority and not just the extreme poles represented in the media.
- Expect the media to deliver bad news and forge your defenses accordingly.
- Learn to read graphs and data critically.
- Calculate risks according to data, not just fears.
- Put things in proportion.

- Question your categories and avoid generalizing or stereotyping.
- Acknowledge that slow change is still change (and collect examples).
- Resist pointing fingers and playing the blame game.
- Resist urgency by taking things slowly.

Yes, *in general* climate change is contributing to a terrible sixth extinction. But overall societal well-being is as good as it's ever been. These two "truths" don't cancel each other out, and of course, one could argue that the former is a result of the latter. There are lots of gray areas. The goal is not to paper over our feelings of fear and despair with a rosy perspective. Rather, we need to recognize that there's complexity and ambiguity in the world—good stuff and bad stuff. It can feel irrational or indulgent to turn away from the bad things that are happening. But it is possible to accept that some things are getting better while also imagining how to address the world's intractable problems. Acknowledging the successes is necessary in order to identify where to devote our energies.

The Progressive Narrative and the Declension Narrative

Critical thinking can also show us the ways in which stories, rather than being entirely objective or "true," are filtered through our lenses and agendas. We can tell a story of the past as bending toward present and future progress or as bending toward decline. William Cronon, an environmental historian, illustrates this in an essay called "The Place for Stories," in which he compares two narratives about the Dust Bowl. Both authors address the same topic and use similar facts and historical sources, but they tell two very different tales. One is about how the Dust Bowl exposed society's weaknesses but brought people together to improve technology—a story of social progress, or as Cronon calls it, a "progressive narrative." The other story, also seemingly objective in its detailing of cause and effect, tells a "declension narrative": the Dust Bowl as an example of how nature avenged humanity's misguided hubris in trying to control it.

A declension narrative operates a bit like a jeremiad, which is a speech or diatribe that laments a decline while heralding a national dream. It berates people for some moral failure and implores them to improve their behavior so that they may obtain the dream. The jeremiad form, a long mournful list of woes, is very common in American political discourse and gains much of its cultural currency from its roots in the biblical apocalyptic tradition. The declension mode is consistent with much discourse in American culture, and it's therefore ready to plug-and-play about any topic du jour, from immigration to abstinence to sustainability.

The template is basically, "Here's the current deplorable state of affairs, which is an insult to our identity; here's how we got there; here's what you need to do to fix it." Let's look at two examples:

Make America Great Again. (Read: things have been getting bad. Our national identity is under threat. We'd better do something to fix it.)

Or,

Climate Change. (Read: things have been getting bad. Our global existence is under threat. We'd better do something to fix it.)

These messages deploy the same template and come from the same rhetorical tradition. They both tap into the threat perceptions of different groups and exploit those perceptions in a feedback loop. There are clear political and moral implications to both the declension and progressive narratives of history. The declension storyline is not "true" or inevitable or objective. It's constructed, and therefore can be challenged or changed. Before you react impulsively to reports that immigrants are threatening national security, or that scientists have discovered that endocrine disrupters are radically altering human biological development, think about how these narratives are designed to manipulate you.

The same goes for the progressive narratives of Steven Pinker, Hans Rosling, and Gregg Easterbrook. In *Enlightenment Now,* Pinker mounts a case for optimism and rejects the "eco-pessimists" of the world. He explicitly describes the ways that their negativity alienates many potential allies. Rosling, in *Factfulness,* is optimistic that the world is becoming a safer, more just place. Easterbrook, in *It's Better Than It Looks,* says our problems are less bad and more fixable than we think. On the one hand, progressive narratives can reassure us that "enlightenment" will give us a reason to live, that the arrow of progress will continue to point up, and that the problems of the past are over (when they might not be). These optimistic analyses can comfort us that all is well and there's nothing we need to do. Thus critics of the progressive narrative worry that it undermines the urgency of the declension narrative, effectively muffling the call to arms.

On the other hand, because anxiety is an ineffective way to prompt action on climate change, a focus on how measurements of human well-being are improving may open *more* doors of action. We can see the benefit of progressive narratives as a counter to the alienating, fear-mongering, misanthropic, and often immobilizing mode of declension. We can also see that both narrative structures reflect the interests of their storytellers, not "truth" per se. The bottom line is that it is more important to debate the emotional merits of progressive and declension narratives than their respective truths.

Growing Your Radical Eco-Imagination

Kari Norgaard calls our ability to see the relationships between our actions and their impacts on Earth's biophysical system our "ecological imagination." Because so much of what we read and hear is toxic, we need to develop a better filter for the stories we digest—to go on what some call a "media diet." Consume stories that keep you focused on the big issues—the fate of our democracy, voter suppression, tribalism, authoritarianism, education policy, and censorship, all of which profoundly impact whether and how we can address climate change and environmental injustice. Deliberate on and analyze the news you ingest before deciding on your next steps. Scrutinize the messages you receive and use critical thinking to ask yourself, Whose interest does this argument serve? Who does it demonize? What is the evidence, and who funded it? What are these images supposed to make me feel? What's being left out? The same cues that prompt us to be loyal consumers of media—alarm and fear—make us surrender our imaginations in favor of filling our days reacting to every stimulus, or worse, numbing ourselves with quick comforts like swiping, clicking, and buying stuff.

It is also important to give yourself breaks from the news. Media can be addictive, and its negativity bias has been shown to be destructive to mental health. Your attention is a resource that you should guard and spend carefully. Balancing how much and which media you consume requires discipline, self-awareness, and maturity. It's like trying to keep your eye on the horizon while a circus carries on in the foreground.

Emotional intelligence is about assessing risks reasonably, which means critically analyzing news stories and recognizing the baggage of negative feelings they conjure up. A healthy dose of perspective will help us take back control of our attention and imagination, so we can operate from an orientation of abundance and spaciousness instead of panic and scarcity.

Mindfulness gives us space to pause between consuming a warning and reacting to it. It also can lead us to pause before consuming messages in the first place, recognizing that they work on our imaginations in potentially destructive ways. Those narratives are not just ambient noise in the background of our psyches and lives. They *create* the affective ecosystem of our daily thoughts and actions. They tell us whether we should be afraid of something, or whether we are okay.

It's not only news outlets and social media that vie for our clicks. Companies and politicians also pay a lot of money to occupy our attention. Your attention is like gold to them. For example, companies may convey a false impression of being more sustainable than they in fact are (a practice known as "greenwashing"), using impossibly alluring idealizations to attract consumers. In the area of cosmetics, for example, women are told that buying natural products will make them beautiful and, similarly, that natural beauty is a sign of a woman's ecological virtue. Privileged standards of beauty are thus aligned with environmental purity, though in fact consuming beauty products grows the GDP first and foremost, while helping neither women nor the environment. Our fears and insecurities—our risk perception—are studied closely and exploited by entities that rarely have our interests in mind. It benefits them to keep us in a state of fear, distress, and need; but it is terrible for us and, I would argue, for the planet.

A statement on the wall at the entrance to one of my local co-ops reads, "Vote with your fork." Making smart choices at the grocery store—shopping locally, buying less meat, purchasing products with minimal packaging—is one way to counter climate change. But don't limit your imagination to individual consumption-based strategies. Other, public actions that shape culture or cultivate collective resilience, such as voting, community organizing, or attending local town halls, expand your sphere of influence beyond the market and can have far greater impacts on politics.

Also examine the contexts in which you experience negative feelings. When I learn how my thoughts, beliefs, values, and feelings are manipulated or privatized by others (through news or other narratives), I am able to observe my feelings from the outside and decide whether and how to act on them.

The process of thinking through our emotions with tried-and-tested tools such as mindfulness and critical thinking might also be called "wisdom." Spend some time thinking about how your emotions can empower you to do the work that needs to be done in the world. What emotions do you need to feel in order to wake up in the morning feeling generous and compassionate? How can pleasure, joy, humor, play, wonder, hope, and optimism be harnessed with sadness, guilt, and mourning as powerful defenses against the largest problem of our time? More important than whether or not Pinker, Rosling, and Easterbrook are correct in their optimism is the way their ideas *make us feel better* about moving forward. We need not believe that the arc of history bends toward justice, nor that we are doomed, to feel that improving the world is worth the effort.

We may feel small in comparison to the mountain in front of us. But if there is power in our emotions, then cultivating emotional intelligence to approach the climate crisis should be a priority. Given how much information we consume, willingly or not, we need the intellectual tools to gain distance from the emotional effects of the stories that circulate in our midst. This will help us discern which causes are worth fighting for, and how intensely we should get worked up about them.

Creating Our Own Stories

Several years ago, I attended a workshop at the Student Leadership Institute for Climate Resilience to learn how to train students in sustainability leadership. One module brought home the ways in which messages enter our bodies and minds; called "Three Stories of Now," it draws on Joanna Macy and Chris Johnstone's "Three Stories of Our Time" from the book *Active Hope*. In this exercise, students are asked to position their bodies to show how three different stories make them feel. The first story, "Business as Usual," recounts the narrative that nothing needs to change in order to address climate change. People interpreted this story gesturally by putting up their hands as blinders, pretending to be smoking a cigar, or folding themselves into the fetal position. The second story, "The Great Unraveling," tells the narrative that the world is moving faster and faster toward chaos and doom; war, climate change, social decline, and suffering are closing in. Here, participants held imaginary guns, distorted their faces in fear or anger, or closed their palms in prayer, looking up to the sky.

The third story, "The Great Turning," is an uplifting tale of humanity being on the brink of radical environmental and collective change for justice and peace. In response to this story, participants moved their bodies into positions showing generosity and connection, holding hands, smiling, seeking contact. In contrast to the stories of business as usual and things falling apart, the Great Turning had the effect of making people feel like wonderful things were on the horizon. The "Three Stories of Now" exercise helps us see and feel the ways that stories affect us. As Macy and Johnstone put it, "When we find a good story and fully give ourselves to it, that story can act through us, breathing life into everything we do."

Marissa, a student from California's Central Valley, grew up in an environment that suffered from severe air pollution and chronic exposure to pesticides. She saw little evidence that anybody cared about doing anything to address these problems. Her sense that things were falling apart around her was amplified by an impression of global deterioration as she read news reports about climate refugees, deforestation, the plight of orangutans, and rising sea levels. She had a strong foreboding that the Great Unraveling was afoot.

In college, her Global Awareness class was assigned to do a research report on the successes of a contemporary social movement of her choice. Everybody shared their reports in public poster presentations at a campus research fair. To Marissa's surprise, the collected stories about so many movements that were challenging the status quo, solving problems, and improving small corners of life buoyed her. Her classmates experienced the same boost in their spirits, as they all realized they were not alone in their worries. They talked about other actions they could do and then banded together to create a garden on campus, which helped with student food insecurity and also replaced an ecologically inappropriate lawn.

The Great Turning, as summarized by Joanna Macy, "involves the transition from a doomed economy of industrial growth to a life-sustaining society committed to the recovery of our world. This transition is already well under way." Transitioning toward something better is a progressive narrative that acknowledges all the work that is already happening. The climate movement is growing: as of April 2019, 62 percent of Americans were at least "somewhat worried" about global warming (23 percent were "very worried"), and 55 percent also believed that it's anthropogenic (human-caused). Both of these numbers are on the rise since 2015. These are the highest levels since 2008, when the Yale Program on Climate Change Communication began administering these surveys and publishing an annual report, *Climate Change in the American Mind*. In the sense that people are mobilizing their energies, these are promising numbers, and they are even more encouraging as regards Gen Z. One poll found that over 70 percent of Gen Zers agree that climate change is a problem, and 66 percent of those think it is a "crisis and demands urgent action."

The Great Turning is a progressive narrative that insists on the power of story to direct our affects and actions. We can choose which story we want to operate from, disregarding those that we know are not true. We can choose which facts to focus on, and in addition, we can choose which story they tell, and how we want to feel about it (being wary, of course, of our confirmation bias, which would lead us to pick only stories that reinforce our existing view of the world). A reasonable approach to media consumption would be to acknowledge more news of collective good things, and exercise skepticism of how facts are manipulated. Don't forfeit your agency by selecting narratives and behaviors that make you feel powerless.

One of the most important insights from progressive narratives is the idea that we are not alone. The more we meditate on our own well-being, the more we realize how tied it is to the well-being of others. Paradoxically, self-study alleviates feelings of isolation, competitiveness, and obsession with our own happiness. As we will explore in the next chapter, the way to enlightenment is through compassion—for self and for others. By seeing our labor in connection with others, we also shed the worry that we're the only ones working hard, and that only we can fix the problems.

As we go about in the world, we can make choices about the stories we create and repeat. We may innocently think we're just reporting the "truth" when we bemoan the environmental crises of our time. But think about the ears that are hearing our words—what does the body they are attached to want to do? Likely, it wants to curl up in the fetal position and shut down to the world. Our attempts to "wake up" our audiences by amplifying this bleak narrative may backfire. Instead, let's amplify stories that expand our sense of abundance to manifest climate justice.

UPPING YOUR EMOTIONAL RESILIENCE MEANS ▶▶
Consuming and Creating Stories Mindfully

☐ Cultivate a radical ecological imagination.

☐ Resist the drain of negativity bias: stop peddling in misanthropy and doom.

☐ Learn how to analyze stories: who is telling them, and why?

☐ Go on a media diet, and get your fuel from solutions journalism.

☐ It's not "the end" of the world unless we resign ourselves to it. Reclaim this moment as "the beginning."

5 Be Less Right and More in Relation

When we wonder why climate deniers refuse to believe the scientific consensus, or ask with disgust, "What is *wrong* with these people?," we are forgetting that most of our beliefs have nothing to do with science and everything to do with emotions. And we are shutting down constructive dialogue before it has even begun.

The battle over climate change has divided us into camps, each with our own interests and concerns about how the current crisis will affect us. Some on the right think climate change will take away their jobs and prompt overreach by an exploitative and incompetent government. Some progressives worry that those who control public discourse will dictate what counts as scientific "truth." Some civil liberties groups fear that the panic incited by climate change will be used to justify isolationist and nativist reactions, like the militarization of borders or the curtailing of civil rights. Working-class populists may see climate change as an elite issue promoted by highly educated scientists who use a specialized vocabulary and championed by privileged people with full bellies who live in nice neighborhoods. Economically vulnerable communities may be concerned less with melting glaciers than with racial profiling or the prison industrial complex.

If these groups can't work together, it will be that much harder to create solutions and to thrive together in a climate-altered world. To effectively address climate change, we need to move away from the perception that the climate movement belongs to the left or the privileged or the powerful.

This is necessary not only in terms of addressing the immediate crisis, but also to maintain our personal resilience, since continually being frustrated by the positions of others will only sap our energy. Paying attention to our emotions helps us transcend our urge to retreat into positions of comfort and opens us up to understanding others' viewpoints. Attention to others' emotions increases compassion, which activates the part of the brain that creates positive feelings. (Empathy, in contrast, can activate a part of the brain connected to pain and distress, so it is less desirable to cultivate, as we will see below.) In the next section, I talk about ways that compassionate communication and a commitment to social justice can help circumvent climate change polarization.

Being Right Is Overrated—Being Heard Is Key

How can we build bridges between deeply entrenched identity categories such as liberals and conservatives, blue- and white-collar workers, coastal dwellers and heartland denizens, urban and rural Americans? Or, instead of seeing climate change as a battle between two sides, perhaps we should be asking what opportunities for alliances and healing engagement with climate change might initiate.

I had noticed that my students were becoming increasingly alienated from certain family members and friends as they became more strident in their thinking about environmental and climate justice. When they went home during school breaks, they frequently engaged in heated political debates with their families and friends. To help them have fruitful exchanges without shutting others down or feeling that they themselves needed to escape, I developed a "Curious, Compassionate Conversation" assignment that gave students tools, such as active listening and nonviolent communication, for holding conversations with people about topics on which they disagree. I wanted to show them how to replace "being right" with building trust and relationship, a skill that is atrophying in us all in this moment of clickism, confirmation bias bubbles, and online conveniences.

Meet People Where They Are

When we talk to people with curiosity, flexibility, and respect, in a way that shows that we understand where they're coming from, they are more likely to respond positively to what we have to say. Climate deniers are neither irrational nor dumb, nor even necessarily anti-science. Instead, their resistance is often connected to feeling that their way of life and their values are threatened. Don't assume that people will oppose the notion of climate change just because they are libertarians or pro-lifers.

Many conservatives care about the environment but hate federal regulation, are suspicious of scientific experts, and think that climate change is a way for so-called liberal elites to gain power over their lives. It's not that they ignore science; they are simply worried that science is being deployed for ulterior purposes that seek to oppress them. Climate activists can build alliances by finding the truth and logic in their concerns.

Letting go of expectations can reveal surprising connections. The Pentagon and insurance companies are unlikely bedfellows with the climate movement, having developed some of the most comprehensive climate change policies and plans in the country. It might also startle you to learn that in 2015 the National Association of Evangelicals recognized climate change as a threat to "the lives and well-being of poor and vulnerable people" and called on its 43 million members across the United States to take action. Think how you can leverage at least some of that collective power by putting aside ideological differences with these groups in favor of agreement around climate change.

Climate scientist and communication expert Katharine Hayhoe explains how this works: "Getting more people on board will not just be about them getting 'enlightened' to our way of thinking; it will involve *them* determining how the problems are framed and approached in the first place. It must be reciprocal if it's going to be lasting. If you don't know what the values are that someone has, have a conversation, get to know them, figure out what makes them tick. And then once we have, all we have to do is connect the dots between the values they already have and why they would care about a changing climate."

Focus on the Local

Exploring the effects of climate change on a local rather than national or global scale also reduces polarization. Law professor Dan Kahan points out that people are more likely to cooperate on actions when they "engage the issue of climate change not as members of warring cultural factions but as property owners, resource consumers, insurance policy holders, and tax payers," or any other type of shared identity. This in turn can have the bottom-up effect of helping communities create solutions together, whereupon they can make demands on their politicians at the national level.

Home in on connections between climate change and the existing, stated concerns of a particular population. Turn your attention to "the places we have loved and lost," as environmental thinker Charles Eisenstein puts it, "the places that are sick and dying, the places we care about that are tangible and experientially known and real to us." Climate change resonates with people when it is *felt* or *perceptible,* either in the body (as with extreme heat or climate-borne illness and disease) or in material life (as with damage from severe weather events). Instead of trying to get people to care about melting ice caps, find out what they care about in their immediate vicinity. Then connect the dots *with* them.

You may cry about the albatrosses and whales that are dying with stomachs full of plastic, but don't expect everyone to get behind climate activism for these same reasons. And don't be upset when they don't. They're just understandably suspicious that the plight of albatrosses thousands of miles away is getting more attention than their own troubles.

Reframe the Issues

One of the most effective measures we can take to promote collaboration across party and demographic lines is to reformulate climate change as a unifying issue or as part of a background story that resonates with the daily experiences of people's lives. Environmental justice scholar David N. Pellow says that the best way to make climate change relevant is by reframing it as a question of health, thus positioning it in terms of the everyday. Health is a bipartisan frame that can make climate change matter across existing political blocks. "What makes lives worth living is more about people, our love of family, hometown, friends, and children.

And their health in particular," says Per Espen Stoknes. "Drawing awareness to the human health impacts," as opposed to the plight of polar bears, bees, or redwood trees, "seems to be an effective method for elevating public concern in the United States," making it a "personal concern for everyone."

Stoknes cautions against reliance on themes of disaster, sacrifice, and uncertainty; instead, he suggests, we should "shift the balance" toward preparedness, values, and opportunity. These frames help align climate change with the top priorities of most US citizens: the economy, health care, and jobs. "What is an effective climate message?" asks Stoknes. It is a message, he says, that is personal, resists the impulse to debate the science and be "right," and attends to people's existing concerns.

Sociologist Arlie Hochschild discovered that shifting the frame was essential to creating a meeting of the minds between her and the mostly white communities she spent time with in Louisiana, which have high levels of industrial pollution and, mystifyingly (to her and climate progressives), high levels of resistance to government regulation of the polluters. She calls this "the Great Paradox." These communities were much more open to talking about environmental issues when they were discussed using symbols or terms that resonated with their deeply held core beliefs, such as "moral purity" or "stewardship." In her research on some members of the Green Tea Party (the environmentally focused section of the Tea Party) of rural Louisiana, framing climate change as an issue of "freedom" or "rights" was much more effective than framing it in terms of sea-level rise, greedy corporations, or scientific data—the issues that most compel liberals and people experiencing the immediate effects of environmental racism, or frontline communities.

Avoid Polarizing Language

You might decide to address climate change by aiding people displaced by political unrest, persecution, rising sea levels, or extreme weather events. You can team up with others to support them without ever using the words *climate refugee*.

In another example, a *New York Times* profile describes how a fourth-generation grain farmer in Kansas has responded to climate change-induced drought by embracing an "environmentally conscious way of farming that guards against soil erosion and conserves precious water. He can talk for hours about carbon sequestration—the trapping of global-warming-causing gases in plant life and in the soil—or the science of the beneficial microbes that enrich his land. In short, he is a climate change realist. Just don't expect him to utter the words 'climate change.'" Miriam Horn, the author of *Rancher, Farmer, Fisherman: Conservation Heroes of the American Heartland,* says that "people are all talking about it, without talking about it. It's become such a charged topic." She challenges the notion that Middle America doesn't care about the environment. "The understanding of our interdependence with nature is not limited to liberal, elitist enclaves. Quite the contrary: The people out there on the tractors and barges and fishing boats understand it as well as anyone, and know that long-term survival requires a changed relationship with nature—not as something to be dominated or exploited, but as an ally of immense mystery and power."

Be willing to keep an open mind. "Purity politics," which Nicole Seymour describes in *Bad Environmentalism* as a didactic kind of eco-fundamentalism, does harm. Working or talking only with people who agree with you on all your top issues will cause you to miss many opportunities to make progress in areas of shared interest. We don't have to get everybody to believe in climate change; we only need to find common ground on which to work together on an issue in which we each have a stake.

Foster Climate Justice

Why focus on climate change when there are so many more immediate, local problems to worry about? Is this something that only those who don't have daily constraints on their health and choices can afford to do? Well, yes, in the sense that a climate strategy that ignores social justice cannot be expected to appeal to people focused on survival, but also no, in that social movements are not limited to the wealthy and the privileged. Social action is essential to everyone's dignity and, in the case of climate change, to their emotional as well as material survival. Climate change is connected to urgent, material concerns of daily life. To the extent that we can enhance the visibility of those connections, rather than divert attention from material inequalities, then we will be more successful at advancing the movement for climate *justice,* not just climate *change.*

One reason climate change has been viewed as an elite issue is that scientific discoveries and concerns can seem far removed from many people's daily lives. When environmentalists claim that climate change is the greatest threat facing humankind, says Mary Annaïse Heglar of the Natural Resources Defense Council, they commit "existential exceptionalism" by not allowing that many communities have endured and are still enduring serious existential threats already. Ignoring their experience "divorces the environmental movement from a much bigger 'arc of history.'" The environmental justice movement is a successful attempt to make environmental issues meaningful to people who have experienced historical oppression.

The Climate Justice Alliance (CJA) is an example of a program that views the climate movement through the lenses of social and racial justice. On December 10, 2018, the CJA posted a press release about what was then a nascent Green New Deal. In the statement, titled "A Green New Deal Must Be Rooted in a Just Transition for Workers and Communities Most Impacted by Climate Change," the CJA pressed for the Green New Deal to mobilize political energy around climate change at the "grasstops" and to include support at the grassroots, including community organizing, collective bargaining and workers' rights, indigenous consent for green projects, a rejection of geoengineering, and assurance that processes would be transparent and inclusive. In February 2019, Representative Alexandria Ocasio-Cortez and Senator Ed Markey introduced a resolution for a Green New Deal that has a deep dedication to climate justice. The resolution not only seeks to make the US economy carbon neutral, create jobs, invest in infrastructure, and provide clean air and water and healthy food, but it also pledges to stop the oppression of "indigenous peoples, communities of color, migrant communities, deindustrialized communities, depopulated rural communities, the poor, low-income workers, women, the elderly, the unhoused, people with disabilities, and youth." Ocasio-Cortez's impassioned speech on climate change to the House GOP in March 2019 illustrates a generational shift toward *social* justice, not just justice for the environment. The social justice check on the climate movement is not just about the "trickle down" of ecological goods that comes with saving the global ecosystem; it's about an entirely different way of understanding the interconnectedness of social and ecological forms of degradation.

As we shed light on the suffering that climate change is causing, it is critical to pay attention to the social, political, and economic structures that exacerbate climate change. Climate change is caused by human forces that can be held accountable. The degree of suffering caused by rising heat, for example, depends on social structures created by humans. We cannot turn down the temperature of the sun, but we can demand responses, such as improvements in health care, worker rights, and infrastructure, not to mention reduction in emissions that result in temperatures rising in the first place. Power over nature, not nature itself, is what creates suffering. Human structures should be seen as a source of productive energy, deployable to address some of our environmental problems. In other words, pay as much attention to voter rights as you do to single-use plastics.

As the Green New Deal and 2019 student strikes show, the rise of youth activism suggests the imminent arrival of a generation that doesn't feel defeated. The climate movement, largely led by privileged white people, is making it a priority to diversify. The alliance between the climate and social justice movements recognizes the potential to bring people together over issues such as labor, indigeneity, immigration, abolition, and demilitarization. We can develop effective strategies by making climate change a social justice issue and recognizing that our position in society—our relative access to power and privilege—affects the way we frame and feel about these issues.

Beyond Empathy

Many in the environmental and climate movements see empathy as the solution to overcoming our disgust and disdain for antienvironmentalists. Empathy puts us in another's emotional shoes and helps us understand how they feel. What could be wrong with that?

In *Against Empathy: The Case for Rational Compassion,* Yale psychologist Paul Bloom explains: "It is because of empathy that citizens of a country can be transfixed by a girl stuck in a well and largely indifferent to climate change." Empathy casts a "spotlight" on individual cases of suffering, often at the cost of alleviating suffering writ large. "Empathy," he writes, "is biased; we are more prone to feel empathy for attractive people and for those who look like us or share our ethnic or national background. And empathy is narrow; it connects us to particular individuals, real or imagined, but is insensitive to numerical differences and statistical data." We don't empathize with those we cannot identify with or see.

Bloom contends that "emotional," as opposed to "cognitive," empathy exploits rather than overcomes our biases by distorting our moral judgments. For example, he sees empathy as being central to both Nazi ideology (overempathizing with the in-group) and concern over mass shootings in the United States (empathizing with victims of an event that is horrible but that represents only 2 percent of gun killings). There isn't anything morally superior about empathy; it may lead to pro-social behavior for some groups, but it can do so at a cost to other groups. We don't need to feel others' pain in order to find good reasons to want to reduce suffering.

And hyper-empathy about others' suffering can get in the way of our ability to function. I can't even say the words *plastic* and *albatross* in the same sentence around some climate activists without triggering the waterworks. A rational (and also gendered) analysis rejects the argument that what we need to do is to be *more* sensitive to others' emotional needs. Women are especially conditioned to care about others' feelings, often at the expense of their own. If anything, we may need to be *less* empathetic.

Empathy is also problematic because it can be perceived as condescending to the recipient. To the extent that it does provide inspiration, literary critic Amy Shuman writes in *Other People's Stories,* a critique of empathy, "it is more often for those in the privileged position of empathizer rather than empathized." This is why gestures of empathy are often seen not as stemming from kindness, as they may be intended, but as patronizing. In any case, empathy "rarely changes the circumstances of those who suffer."

The desire to build relationships, especially across ideological and political differences, emerges from a variety of affects, including, but not limited to (and sometimes expressly *not*), empathy. We should not be required to surrender our anger, or insist on a safe stance of harmony, in order to be willing to cooperate. We do not need to love our neighbors; in fact, we may not like them at all. The point is not to love them or empathize with them, but to be generously curious about the feelings that motivate them, for the larger goal of helping society function as well as it can for as many as it can. Understanding and care are possible without empathy.

Furthermore, while understanding others' positions is a step, it does not guarantee cooperation or relationship-building. In that sense, empathy is not an end in itself, even though it is certainly a desirable alternative to more toxic emotions. On the surface it appears to ask that we shortcut thinking and go straight to an emotional appeal to "just get along" for the purpose of some greater good of harmony. But harmony, comfort, and empathy are neither required for, nor the point of, democracy or social change. We can cooperate and build community—indeed, we *must* learn to be able to—*despite* anger and discomfort. That is the real challenge—to hold space for both righteous anger and curious compassion.

Compassion—Self-Preserving and Respectful

Neuroscientists such as Tania Singer have shown that empathizing with someone in distress lights up parts of the brain associated with pain, while compassion lights up parts of the brain associated with love. Empathy involves feeling another person's suffering, as if you are experiencing the same thing. Compassion is the ability to care about another's suffering.

Boundaries are essential to our self-preservation. Yet whereas empathy dissolves our boundaries, compassion maintains them: it merely asks that we remain curious and understanding, and that we desire for others' suffering to end. The term *compassion fatigue*, used to describe the unique exhaustion experienced by doctors, trauma workers, and emergency responders, in fact confuses compassion with empathy. Empathy fatigue is a significant cause of burnout—it can result in withdrawal, and chronic empathic distress can harm your health. Whereas empathy can be exhausted, compassion is renewable from our reserves of abundance.

The word *compassion* comes from the Latin root *compati*, meaning "to suffer with." Mindfulness expert Judson Brewer describes suffering alongside someone as quite different than taking on suffering. But how we can stay connected without becoming exhausted by that connection? One requirement is to have compassion for oneself. Buddhists believe that compassion for the self is the solution to suffering—our own and others'. Without compassion for yourself, it is impossible to have compassion for others. And without compassion for yourself, compassion for others may overcome you. Thus compassion is a crucial tool for becoming resilient for the long haul.

Our energy to care for others comes from our attending to our own needs. The more we are able to nurture our own deepest existential desires—not just superficial wishes, but needs like feeling loved—the more energy we have for others' needs. In order for us to continue to love the world enough to keep working to improve it, we must cultivate compassion, first of all for ourselves.

Compassionate curiosity, by helping us understand the other side's myriad and nuanced positions, can achieve the desired end of building trust enough for cooperation, even in situations that make us angry. For example, when I first heard the chant "Drill, baby, drill," popularized by ex–vice presidential candidate Sarah Palin in her quest to open public lands to oil drilling companies, I was horrified and angry. But I relieved my feelings by exploring the emotional motivations of this rhetoric and its impacts on me. I asked myself, What are my fears, and to what extent are they manufactured and exploited for political gain? What are the chanters afraid of? What are they attached to? What does "drilling" symbolize for Palin and her followers? I assumed the best: they were probably not trying to call forth the end of the world or the destruction of all life, which is how the chant landed on my ears. As long as I accepted that the protesters had humanity, there was a channel open for us to talk.

It doesn't make us feel good to be angry and to imagine that others are ill intentioned and heartless. For entirely selfish reasons, not empathy, we should replace horror with curiosity and compassion. An orientation of curious compassion alleviates suffering, and it also may lead us to find more effective political strategies to change others' hearts and minds. Understanding that fear is at the root of much of our politics enables us to remain calm and make better decisions. What do you fear? Start there.

It is easier to be compassionate about others' views about climate change if we see these other worldviews as varying by degrees in many different ways, rather than clustering at the poles. Hans Rosling describes our tendency to see the world in black and white as a holdover from evolution, what he calls the "gap instinct," which "makes us imagine division where there is just a smooth range, difference where there is convergence, and conflict where there is agreement." Those who produce stories—both fiction and nonfiction—exploit our gap instinct to manufacture drama between good and evil, reinforcing our tendency to see the world this way.

There's ample research showing that the vast majority of Americans do not harbor the extreme opinions that are represented in the media. The notion of the left versus the right is misleading, and we would be better served realizing that there is a great swath of Americans who are *not* polarized and who *are* interested in having robust conversations about controversial topics. In the United States, only 10 percent of Americans adamantly do not believe in climate change, which means there is a huge coalition of the concerned.

It would behoove us to continue to point out issues that concern people across party lines, including the economy, jobs, and health care. Many people in the center and on the right care about climate change. Unlike abortion or guns, bipartisan collaboration on climate change is on the horizon, especially when youth are taken into account. "I don't personally know anyone involved in young, right-of-center politics that doesn't believe climate change is an issue," said a young Republican in an article about the American Conservative Coalition. "We can talk about this. Conservatives that care about the environment do exist."

Raising the pitch of our anger and deepening the divides between ourselves and others only fuels greater fear and anger. This sets off a feedback loop, which will achieve the opposite of mobilizing public will about climate justice. Even if we were to "win" the political debates on these ideas, the victories may last only until the next election. Effective action against climate change takes longer than that.

The endgame of compassion is not ultimately to agree with others or make them agree with us. Rather, it is to recognize how our inability to communicate is destroying the very fabric of democracy. Following philosopher Richard Rorty, we must recognize how "speaking differently"—not in the aggressive ways we see and hear in the media—is more effective than "arguing well" in bringing about social change. We must shift the goal of "winning" our arguments to the far better goal of building shared interests with people whose positions may be different from ours.

We operate much more graciously when we assume that everybody is doing their best. The topics on which we disagree politically may be minor compared to values we have in common. A person who thinks that climate change is a hoax may also be raising her kids to not squander resources, think that the United States should be open to many races and religions, believe that gay marriage is a right, and love chocolate. Can we find a way to build social capital with this person?

If we proceed from the assumption that we may have something in common, and that we have a shared desire to contribute to the improvement of social conditions, we can have a conversation and perhaps even build a relationship based on respect for our differences. If we feel we're constantly battling some "evil other," our anger and fear will deplete us.

Knowing that a lot of people are with us, and that the rest are not as malicious as media would have us think, we are much less likely to give up. In the meantime, we may even build relationships across a lot of divides, alliances that are necessary to be effective in whatever we seek to do.

Cultivating Curious Compassion for Others

- ☐ Make climate change resonate by calling it by its many other locally relevant names.

- ☐ Learn how power and activism work in communities unlike your own.

- ☐ Don't dispense partisan or elitist wisdom: ask questions of others, and listen to their answers.

- ☐ Forgo the instant gratification of being "right" in favor of building long-term solidarity across party and class lines.

- ☐ Focus on common values like health, morality, freedom, and beauty to emphasize the universal benefits of climate activism.

- ☐ Cultivate compassionate curiosity and understanding toward others.

- ☐ Start with compassion for yourself.

6 Move Beyond Hope, Ditch Guilt, and Laugh More

Let me share a story of failure. I had a student, Ben, an athlete and fraternity brother, whose balance of earnestness and arrogance made him a great contributor in class. As a white male, he struggled with "being the cause of all the problems we learn about," as he put it, and although I tried to guide him from fragility toward solidarity, he eventually dropped the environmental studies major. I feel like I failed him because at the time I didn't have the tools to help him work through his guilt to become the contributor he had the potential to be.

"Green guilt" plagues many of us, not just white males. When people realize how social injustice spirals from environmental problems, and the extent to which they, by virtue of being in the Global North, are complicit in causing suffering, they can experience eco-guilt: guilt about how our consumer habits are destroying the planet, guilt about how race, class, gender, ability, or zip code can compound suffering, guilt about not doing enough, guilt about prioritizing more immediate tasks, guilt about not showing up to that or the other protest or meeting, guilt about . . . The list goes on.

Some guilt stems from the privilege experienced by those of us with ready access to food, mobility, education, safety, and health. In the Global North, some of us enjoy greater privilege than others, and the process of becoming "woke" may mean reckoning with the shame that comes from realizing how much one benefits simply by being born with a certain set of genes and in a particular place. Being discomfitted by the ways that one's identity grants unearned privilege, especially if one is white, is a step along the journey to solidarity. I want to suggest that if climate justice is your goal, you should ditch the green guilt and eco-fragility, learn to appreciate the nuances of hope, and take time to love, laugh about, and delight in the world and in your work.

Guilt Ain't Green

Experiencing guilt is part of the process of accepting your privilege, but it can trap people in agonizing self-reflection—often dubbed "analysis paralysis." When privileged people get stuck in guilt, they paradoxically can feel excused from the work of justice. The desire to not feel guilt is often greater than the desire to right whatever wrong is making us feel guilty, such as how we've failed the planet. Guilt is self-interested; alleviating the unpleasant feeling of guilt becomes more important than alleviating the suffering of others.

Guilt is thus ineffective for working toward racial justice, although it's used as an affective strategy in most environmentalist messages. Eco-guilt is the feeling that we have so harmed the environment that we are eternally indebted and will never be able to right that debt, no matter how many green choices we make on a daily basis. Eco-guilt makes us feel bad that we are robbing the planet of its past (fossil fuel takes a long time for the Earth to make) to support the quality of life we enjoy now. Our privileges and advantages, some of which are unearned, often become a source of guilt when we start to notice them. Even so, although we didn't originate industrialization, for example, we find it difficult to opt out of using fossil fuels. We may not be colonists, but we may have profited from colonialism. Those of us who are able-bodied benefit from the world being designed to support our ease of movement.

Guilt may initially feel like a form of humility. But "no one likes to feel bad, especially not about themselves," as literary critic Nicole Seymour points out, so although guilt may trigger short-term reactions, it rarely keeps people involved in climate work in the long term. Appeals to eco-guilt can even be distasteful to marginalized communities, particularly if they do not address the specific problems that affect them. Attempts to change consumer habits to alleviate the suffering of sea turtles, for example, have little relevance for people suffering from environmental pollution and related health issues, which disproportionately impact isolated or economically depressed areas.

I've done the "ecological footprint" exercise about five times. This task helps individuals understand the impacts of their consumption habits (what they eat, what they buy, how often they fly in airplanes) in spatial terms that are easy to grasp: in my case, we would need six Earths for everybody on the planet to consume as much as I do. Each time I do this exercise, I feel a wave of guilt before I even start, and I curse the person who is making me do it. Afterward, I make minor capitulations to reduce my footprint, like carrying cutlery around with me so I don't use plastic forks. These changes typically last about three weeks. And I'm not alone. Behavioral adaptations based on unpleasant feelings, like guilt and self-loathing, are not sustainable.

By contrast, I changed my consumer behavior far more after reading Robin Wall Kimmerer's *Braiding Sweetgrass,* for reasons having little to do with guilt. Instead of treating everything we consume as if it had "just fallen off the back of Santa's sleigh," Kimmerer says, we would do well to remember how the objects in our daily life were "ripped from the earth." She calls the rules "that govern our taking, shape our relationships with the natural world, and rein in our tendency to consume" the "Honorable Harvest" and talks about acting in ways that acknowledge our relationships with the living things around us "based on accountability to both the physical and metaphysical worlds. The taking of another life to support your own is far more significant when you recognize the beings who are harvested as persons, nonhuman persons vested with awareness, intelligence, spirit."

She also sees inanimate objects, such as the pen I just bought, as the product of labor and resources. It cost a lot more than the dollars I paid for it. The resources required to make the pen, transport it to the store, and dispose of it after I finish using it amount to far more than the material of the pen itself. The production, transport, and disposal phases of our consumption are hidden from our view, and we expect the costs of those phases to be hidden from our pocketbooks. In seeing these previously invisible costs of the pen as gifts that constitute the pen, I tap into an affect of care, love, and what Kimmerer calls "practical reverence." Thinking about my daily actions in terms of practical reverence, I feel gratitude for every piece of toilet paper I use and for the resources and labor that make my waste invisible to me. I acknowledge that every one of my choices is felt by people and nature somewhere else. After a lifetime of not gardening, I made a compost pile because I liked the feeling of processing at least some of my waste. Care, love, reverence—I experience these things as desirable, pleasurable. I honor the gifts I receive from the Earth, and I do so without guilt, or at least, without guilt as an eclipsing affect. I'm much more compelled by the appeal to sacred abundance than by the guilt-laden appeal to scarcity and sacrifice.

The argument that I should not buy a pen because it might end up in a turtle's stomach may conjure guilty pathos and inspire empathy for the turtle, which in turn draws the turtle out of the deep sea and into my realm of awareness; but ultimately, even images of suffering turtles don't keep me—or others—from buying pens. We doubt that not buying pens will save turtles. This explains why so many people still purchase single-use plastics. They may feel that abstaining from buying plastic cutlery for a picnic is negligible in the broader picture of planetary salvation, and therefore decide that the small scale of personal sacrifice won't matter.

Practice green consumerism because it feels right—because you *wish* to—not because you'll feel guilty if you don't. But don't let green consumerism make you complacent. When you limit your arena of social change to what you buy or don't buy, you can lull yourself into thinking that's enough. Remember that the pollution and resource costs of a pen pale in comparison to the impacts of industries. Although we wield power as consumers, we have greater impacts as citizens, community members, and social-change agents who can do much more than not buy plastic cutlery or pens.

I find myself responding resentfully to environmental messages that ask me to feel guilty. Kimmerer's non-guilt-based pathos helped me overcome my own affective dissonance with respect to guilt. That is the model we should use in encouraging others to become climate activists. "Make justice and liberation feel good," insists adrienne maree brown. Humans will come back over and over again to feel pleasure.

Our intention is our compass; it sets our course. What do we gain from the work we do? If the answer is "alleviation of guilt," we are not likely to return for more. Guilt as a motivator reduces our potential to make the efforts we are capable of; because we keep trying to measure the magnitude of our impact and gauge whether our debts of guilt are paid, we always fall short. It becomes an excuse to stop trying. Instead of guilt, shame, or fear, how can we make love, pleasure, and desire draw us and others to the work of climate justice?

Less Stick, More Carrot

A few years ago, I started to encourage my students to think about their experiences in class as a journey through stages of feelings. One day, one of my seniors joked that she was going through a stage of worrying about her future and grieving about social and environmental injustice that she called the "baking stage." She started bringing her baked goodies to class, and made a reputation for herself as the person who fed us delicious food. The pleasure she brought us was an outlet for her stress. At first, I dismissed this "stage" as the joke it was meant to be, but I came to realize that the baking stage might be necessary for many of us (see the illustration of the affective arc—figure 2 in the introduction). Anarchist feminist Emma Goldman is famous for supposedly having said, "If I can't dance, I don't want to be part of your revolution." If the future we are working to protect doesn't have pleasure, my student was saying, then she didn't want to be part of it.

Make a daily practice of the joys that make life worth living. Do not defer this to when the revolution is over. "Let me remind myself why I'm doing all this," my student's actions said. In the process, she built social capital, warmed everybody's hearts, filled their stomachs, and helped yet another group of seniors walk across the graduation stage.

Experts on grief tell us that to cope with grief in a healthy manner, we must confront it and feel it, not try to avoid it or cover it over. This is true also for ecological grief. Getting more comfortable with negative feelings is essential to building personal resilience. As I argued in chapter 2, suffering is a fact of life, and the sooner we can accept that, the sooner we can detach ourselves from expectations that things should be different. Climate change is not the only kind of suffering that the world has seen; hence all religions and spiritual philosophies have prescriptions for dealing with negative emotions. We can draw on ancient wisdom and psychology to learn how to face these feelings. The inability to cope with negative feelings is one reason rates of depression and anxiety are so high in Generation Z.

Prioritize the pursuit of positive emotions in your advocacy work and in your personal practices. Seek life-affirming activities, some of which may play no obvious role in your advocacy (like baking cookies), and some of which are central to building the movement (like joining a committee or sustainability club). Both "unproductive" and "productive" positive actions, as Seymour describes them, can help us recharge ourselves and others and build resilience, as long as they are not also self-destructive (like social media addiction, which may feel pleasurable but profoundly undermines our ability to replenish ourselves).

Whether our eco-grief is caused by acute forms of trauma, anticipated trauma, or witnessing others' trauma, we need constant reminders of what we are living for. To the extent that we have any choice over the matter (and I want to emphasize that people with clinical depression and anxiety often *do not*), we simply cannot occupy a space of sorrow and heartbreak for long periods of time without great cost to our health and, ultimately, to the causes we care so much about. Spend some mental energy imagining the restitution for your suffering. Take time for practices as seemingly small as pausing to be grateful for rain or as grand as starting a book group to talk about fantasy world-making. Visit with friends, or plant a garden. "Sustain the ones who sustain you and the earth will last forever," says Kimmerer.

Hope Isn't Enough

When dissident Václav Havel was fighting for reform and democratization in Czechoslovakia in the 1960s, he said, "Hope is not the conviction that something will turn out well, but the certainty that something makes sense, regardless of how it turns out."

Havel's approach to hope differs from that of many climate change journalists, scientists, activists, politicians, and educators. From Rebecca Solnit's *Hope in the Dark* to Bill McKibben's *Hope, Human and Wild,* nearly all the usual suspects seem to have received a memo that they should add a spoonful of sugar to help the apocalypse go down (which is perhaps just one reason Jonathan Franzen's 2019 *New Yorker* piece, "What If We Stopped Pretending," which didn't offer much sugar, caused such a stir).

Along with ecocritic Alexa Weik von Mossner, I reject the strategy she describes as the "principle of hope," or hope as an afterthought, which involves "add[ing] a hopeful silver lining to . . . otherwise dystopian tales." The "ending on hope" trope can be found everywhere. From children's environmental television shows to college course syllabi, alarming information is softened by a positive ending, as if to say, "You can do something about this" or "Here is somebody solving this problem; all is not lost." Books, articles, and podcasts claiming to offer hope have become a cottage industry of climate change communication. Swedish youth activist Greta Thunberg even pokes fun at this trope in the TED Talk she did in the wake of the school strikes for climate she precipitated across the globe in 2018: as she concludes her talk she teases her audience for expecting a hopeful note, but she refuses. Instead she warns against holding up hope as the endgame; when action happens, she says, hope will inevitably be everywhere. For Thunberg, though, hope is not the point. Perhaps you even bought this book hoping I might have some hope to peddle. Maybe I do. But my desire is not to leave readers feeling *hopeful*, but to leave you feeling *efficacious*. Feeling like you have the capacity and power to do something, including the most important task—preserving yourself for a lifetime of thriving in a climate-changed world: that's the point. Hope by itself is beside the point.

Derek Jensen, in "Beyond Hope," urges us to "get to work." Like Thunberg, Jensen calls the hope habit a distraction from action: "I don't have much hope. But I think that's a good thing. Hope is what keeps us chained to the system, the conglomerate of people and ideas and ideals that is causing the destruction of the Earth." Ultimately, I agree with Jensen's startling conclusion that "hope is a longing for a future condition over which you have no agency; it means you are essentially powerless."

I am in favor of a complex affective orientation toward climate advocacy that avoids some of the pitfalls caused by a superficial attachment to hope. If we need hope in order to keep ourselves engaged, then we may require tangible evidence of our impact (the "instrumentalism" approach that I rejected in chapter 3), and will flag in our efforts whenever evidence is elusive. This is *utilitarian* hope, and it is not a good strategy.

What we need is *critical* hope, in the words of education expert and activist Jeff Duncan-Andrade. Not "hokey hope"—artificial,

Pollyannaish—but hope that gives people new "resources to deal with the forces that affect their lives" and acknowledges that the "painful path *is* the hopeful path." This needs to be combined with a radical imagination about the future we desire. As von Mossner maintains, hope alone is short-lived unless coupled with desire for "future worlds that are not depressing but instead so enticing that they might lead us to *yearn for a new way of being.*" What's better than hope? Empowerment to work toward a way of being for which you *yearn.*

Desire is an essential affect for building resilience. Desire is "hunger or longing to reach a state that seems attractive, satisfying, and pleasurable," according to von Mossner. The second noble truth of Buddhism states that desire is the cause of our suffering, and so it may seem strange that I advocate for desire here. But as I argued in chapter 2, Buddhism distinguishes between the kinds of desire that are distractions from suffering, but which never alleviate it and in fact often cause further suffering, and the kinds of desire that are necessary to alleviate the experience of suffering. Mindfulness helps us sort out what we truly desire from what we desire as a quick fix to escape from our existential suffering. Hope plus existential desire adds up to a sense of purpose, a reason to withstand the suffering for longer-term fulfillment. Hope alone cannot get us there; indeed, dashed hopes can lead to despair. Desire is needed to sustain hope in the face of adversity.

Desire entails the anticipation of pleasure, whereas most climate narratives invoke threats and risk, as I discussed in chapter 4. The carrot of desire works better than the sticks of guilt or fear. Pleasure is a crucial dimension of desire. Applying Buddhist notions of desire to her theory of pleasure activism, brown stipulates that "the deepest pleasure comes from riding the line between commitment and detachment." Psychological studies show that "reframing climate science in terms of the benefits of making change—as opposed to the dangers of continuing with the status quo" is one method of shifting polarization around climate action. What are the benefits of making change? Going back to the visualization exercise in the introduction, what will it take to imagine thriving in a climate-changed world?

Because of her concern about contemporary politics and mounting anxiety among activists, brown wrote *Pleasure Activism: The Politics of Feeling Good.* In it, she asks how the pursuit of social justice can be the most pleasurable human experience. In *Emergent Strategy,* she also admonishes the all-too-willing acceptance of misery that accompanies so much activism. The culture of activism often demands seriousness, an "if you're not miserable, then you're not paying attention" position. brown rejects this orientation, arguing that we have to stop demonizing pleasure and ennobling misery. She refers to the long history of "misery resistance," practiced particularly in Black communities, "of dancing, singing, and loving *as forms of* resisting, and releasing oppression from the collective Black body." In the next chapter, I explore how joy and pleasure are a kind of resistance to the oppressor's need for our spirits to be deflated so we won't fight for justice. These feelings are not frivolous extras; they are requirements.

Nicole Seymour's *Bad Environmentalism* explores humor, which some eco-psychologists recommend as a "resiliency skill." The paradox of simultaneously accepting death (the planet's and ours) and cherishing the planet and ourselves enough to actively care for both is, Seymour suggests, a necessary kind of *irony.* Irony characterizes much environmentalism these days, for example the notion that we have to keep acting as if our actions matter despite evidence that they may not. But gift economy advocate Charles Eisenstein points out the benefits: "It's as if giving up on saving the world opens us up to doing the things that will save the world."

Accepting this irony, or "learning to die in the Anthropocene," as Iraq war veteran and journalist Roy Scranton puts it in his book of the same title, is perhaps a better alternative to hope. Acceptance, not hope, is the remedy for nihilism, Scranton says. The ability to accept the seeming contradiction that the world is dying even while we operate from a place of hope and desire is the aim of spiritual awakening. In Buddhist terms, "learning how to die" and "accepting impermanence" are not as morbid as they sound.

On the contrary, acceptance of mortality enables us, ironically, to cherish life. Or, as Scranton puts it, "It's at just this moment of crisis that our human drive to make meaning reappears as our only salvation . . . if we're willing to reflect consciously on the ways we make life meaningful—on how we decide what is good, what our goals are, what's worth living or dying for, and what we do every day, day to day, and how we do it."

In *The End of Ice: Bearing Witness and Finding Meaning in the Path of Climate Disruption,* journalist Dahr Jamail writes about grief as a necessity for honoring suffering. Grief is the opposite of hope, but, perhaps counterintuitively, it's a more important affect for doing the existential work of facing climate disruption. Whereas hope is often sought as a distraction from the gravity of what's really going on, for Jamail, "a willingness to live without hope allows me to accept the heartbreaking truth of our situation, however calamitous it is. Grieving for what is happening to the planet also now brings me gratitude for the smallest, most mundane things. Grief is also a way to honor what we are losing. . . . My acceptance of our probable decline opens into a more intimate and heartfelt union with life itself." Grief allows Jamail to "fall in love with the Earth in a way I never thought possible." Embracing grief and decline is not a morose, fatalistic orientation. Facing death— the planet's and ours—opens us up to the love that is necessary in order to sustain the work of climate action. Punishing ourselves for being a part of the humanity that has caused so much destruction can activate a savior complex or a politics of sacrifice, and assumes that the goal is moral purity. It's preferable to aim for humility instead, and to accept our own entanglement in systems of injustice.

This is the point where grief studies, Buddhist mindfulness, and social movement strategy come together. We are of no use to the cause if we are burned out by apocalypse fatigue or so over-whelmed by fear, shame, and sadness that we simply opt out to save ourselves from pain. Even if we are skeptical of Pinker's and Rosling's progressive narratives that the world is better off than it ever has been, it behooves us to focus on a wider variety of facts than we might get from the mainstream media. We must maintain our motivation to keep *desiring* the future, and keep working to cre-ate the conditions that will make that future not just tolerable, but worth living for.

Find beauty, savor the small gifts of being alive, see everything you possibly can through the lens of being blessed rather than vic-timized, recalibrate your efforts toward the small and local, collect and create positive stories, heed your calling by not trying to be more than you are, take yourself less seriously, and pause to inhale deeply and honor the moment. This is what it means to learn how to die in the Anthropocene. It's a Buddhist or existentialist goal that means sustaining your will to live so that you can keep trying to stave off the end of the world.

Environmental ethicists ask, "How can we get people to care more about the environment?" If we really care about addressing climate change and want to get others to do the same, we must make the work desirable for ourselves and for the sake of bringing more people on board. One way we can do this is by accepting the impermanence of the world and of ourselves. What do we have to lose by taking pleasure in our work, slowing down, and doing the part we *can* do, *well?*

UPPING YOUR EMOTIONAL RESILIENCE MEANS ▶ ▶

Ditching Guilt, Moving Beyond Hope, and Laughing and Loving More

- ☐ Examine guilt as the residue of self-aggrandizement, savior complex, or privilege.

- ☐ Don't worry about creating hope. It is a natural byproduct of the work.

- ☐ Harness the feelings that generate long-term commitment, such as love, reverence, or pleasure.

- ☐ Examine your deepest desires, below and beyond the dopamine-hits of addictions and distractions: they will sustain you much longer.

- ☐ Embrace irony and impermanence.

- ☐ Live in the world you want to save. Create time and space for wonder, leisure, and play.

7   Resist Burnout

adrienne maree brown dedicates her book *Emergent Strategy* to activist Grace Lee Boggs, who "opened the door to emergence and pushed me through, who taught me to keep listening and learning and having conversations. She said, 'Transform yourself to transform the world.'" brown urges us to learn to be collaborative, to trust others, and to tell stories as ways to organize "based in love and care rather than burnout and competition."

The first place to practice loving-kindness is with ourselves. Unfortunately, those of us who feel the most passionate about social change are also the most likely to fail to be considerate of ourselves. It's easy enough to read a list of to-do tasks for self-care, but it is an entirely different challenge to implement them in our lives. Ask yourself what barriers keep you from doing it. We assume that "the cause" we are fighting for can be achieved only if we sacrifice ourselves. Isn't that the heroic narrative of change? The ethos of martyrdom is seductive in any social movement.

I came to recognize my own attachment to this logic when I became a mother in my mid-thirties. No level of sacrifice, I thought, would be sufficient to prove my commitment to motherhood. I struggled to find the limits of my capacity to serve my kids and my family, and to know where to place boundaries to preserve myself. I felt that everybody else's desires were more important than my own. Compared to my children's needs to be shown love or a desperate student's need for solace after class, *my* need to get an hour of exercise or a full night's sleep always seemed too selfish and infinitely deferrable.

Yet I came to deeply resent this pressure. I became self-destructive and my nerves began to fray. At that point I realized that only if I protected myself and my time would I have the energy to devote to my family, my students, my advocacy work, and my self-care. The first hurdle I had to overcome was my assumption that the needs of others mattered more than my own—right up to and including small, seemingly minor daily practices, like making my bed.

The second hurdle was my assumption that the climate required my self-sacrifice. Environmentalists are particularly prone to such martyrdom. The planet seems to demand that we pay our debt to it by killing ourselves. Environmentalist doctrine preaches that pleasure, especially consumer pleasure, is experienced only at the cost of biodiversity, ecosystem services, and environmental justice. We can thank colonial expansion for every taste of chocolate we enjoy, which now comes at the additional expense of virgin oil palm forests and the orangutans that live there. Every moment we spend breathing deeply is a moment that could be spent protesting the siting of an incinerator in our neighborhood or the rollback of auto emission standards. For many of us, this level of engagement feels like a matter of life and death: who has time to sleep when so much is under threat?

Those of us who are not on these frontlines may feel that we owe a debt of sacrifice for the pleasures of our relative comfort, that self-imposed suffering goes some way toward demonstrating solidarity with or honoring those who *are* suffering. For these reasons—call it ego-guilt—self-care may feel like an indulgence that we have no right to.

Certainly, the capacity to take care of ourselves relates to the perceived seriousness of our personal or cultural survival. Deciding whether the cause will fall apart if we take a nap is highly personal. But I would encourage all of us to remember civil rights activist Audre Lorde's words when she was fighting cancer: "Caring for myself is not self-indulgence, it is self-preservation, and that is an act of political warfare." Or, as education coach Elena Aguilar writes in *Onward: Cultivating Emotional Resilience in Educators,* "In order to create the just and equitable society that I know so many of us yearn for, we need tremendous reserves of resilience." To "change the macro conditions in which we live and work," she concludes, "we'll need all the physical and emotional resources we can muster."

It was only when I discovered research on how depression in mothers affects their children that I accepted the possibility that any action I could take to fend off depression was good not only for me, but for others—the ones I was sacrificing myself for in the first place. If I really cared about my kids, I would have to put myself first occasionally. I did some self-study, trying to understand why I felt I was a good mother only when I was sacrificing myself. What was the root of my guilt when I pursued "selfish" things like writing, researching, taking a walk, or spending time with a friend? What was I afraid of?

Once I found the answers to these questions, I was able to stop the sabotaging behaviors that took me away from myself and from the public work I otherwise care so much about. I realized that making my bed was a sacred act of self-preservation, one that set my posture toward a day of caring for others, and that saying no to one demand might open space for a yes that would serve me *and* the movement. Aligning one's time and energy with one's priorities is a lifelong practice. Yet it is the only way to resist burnout, which is what the people, species, and ecosystems we care for need us to do.

In researching the relationship between self-care and activism, I came across books, blogs, and podcasts, such as Van Jones's "Breaking Out of Our Resistance Bubbles," that argued for a balance between facing inward and facing outward. The message is that too much facing inward can lead to a retreat from the work in the world that needs to be done, while too much facing outward can lead to burnout.

In these analyses, curiously, meditation was often criticized as apolitical "navel-gazing," at least if it's all a person does. But this reputation emerges from a misunderstanding of the purpose of mindfulness. As a long tradition of social activists and movement leaders demonstrates, there is no contradiction between self-care—such as a meditation practice—and political engagement. If an oppressor is trying to deplete a person's emotional and physical resources, stripping them of the energy they need to resist oppression, then any behavior that protects those resources likewise counts as resistance. Cultivating pleasure, relationships, and energy through sleep, retreat, silence, inaction, and the pursuit of activities that generate passion and creativity is essential for us to maintain the human capital required to keep engaging in political work. When we judge others for taking care of themselves and putting "the cause" on the backburner, we ignore this reality. Just as the work required to keep children, families, and the elderly fed, rested, and healthy—the care work of social reproduction—is typically ignored in analyses of economic growth, so too the resource of our own psychic health is undervalued in the activist economy of time and energy.

The best way to resist burnout is to commit to a mission statement that helps you triage the demands on your energy, stop performing busyness as a badge of commitment, and cultivate a daily practice of self-care. I do not mean the indulgent, consumerism-based "self-care" that allows you to escape your grief, but rather the kind that helps you heal and prioritize so you can stay in the game.

Many books outline how to do this, and I list some of my favorites in the notes to this chapter. Among the most effective methods of self-care, as covered in previous chapters, you can:

- Practice mindfulness and gratitude.
- Avoid self-sabotaging habits.
- Limit your use of social media.
- Get enough sleep.
- Focus on the tasks you find fulfilling and in line with your priorities.
- Say no to requests that are not aligned with your priorities.
- Foster a support network.
- Cultivate compassion.
- Care for others.
- Visualize good outcomes.
- Create better stories.
- Celebrate successes.
- Seek beauty and pleasure.

Know when to step back, to recuperate for the next surge. "Opting to pause and reflect," says journalist Thomas Friedman, "rather than panic or withdraw, is a necessity. It is not a luxury or a distraction—it is a way to increase the odds that you'll better understand, and engage productively with, the world around you."

If we think of ourselves as acting alone, we feel that there's no time to step out of the fray to recharge ourselves. If we recognize that we are a collective—a choir of voices—we know that the song will continue while we are recovering our breath, and that we can rejoin when we can. How do we maintain this awareness of ourselves in collectivity, rather than trying to fight the world's problems solo? As I outlined in previous chapters, it requires a pursuit of community (often virtual), a rejection of the news cycle that almost never shares stories of success, and an unshakable belief that self-preservation (as opposed to self-sacrifice) is *both the fuel and the purpose of* the movement.

Clinical psychologist Sam Himelstein works with students who have suffered trauma. His definition of self-care encompasses both care of self and care of others. For him, the four key components are:

1. Regular cultivation of relaxation response (3Rs): things like watching TV, going into nature, getting a massage.
2. Effortful training: These are things like more sustained meditation or exercise where the payoff comes over a longer time period.
3. Creativity: something that gives purpose and adds vibrancy to life. Writing, reading, painting, or other passions are examples.
4. Advocacy: everything from learning to say "No" (set boundaries), to working at a higher level to impact policy or structural change.

These components reveal the mutually supportive nature of self-care and advocacy. Advocacy includes both setting boundaries by saying no to external demands and working to address the structural sources of problems. If self-care means advocacy of self, then it must mean conservation of one's energy *and also* dismantling of oppressive structures. Instead of asking, "Should I sleep or should I keep working?" we might ask, "What is the best way I can advocate for myself right now?" The answer may be to show up at a city council meeting, or it may be to spend an hour with a friend. There is no one-size-fits-all prescription; only you can decide what you need in each moment to remain engaged and energized.

If you are reading this book, you are probably already profoundly committed to climate justice. In fact, the vast majority of people who are involved in climate work and in the climate justice movement are convinced that ensuring the survival of the planet will require all the energy we can possibly muster, for a long time. Not many of us need cajoling to care *more* about these issues. We tattoo "Resist!" on our bodies. When I was a participant in the vision-change-action workshop I described in the introduction, I reviewed the list of action items that our group wanted to do, recognized I had already done the majority of those things, and decided that *my* very next step would be to take a nap. I realized that I had achieved a lot but hadn't taken the time to celebrate my successes, much less recover from the effort. My own martyr complex made me feel like the work was never done. The work *is* never done, but that's why we need to realize we are part of a collective effort and that the movement requires that we *not burn out*.

I still struggle with the feeling that the drudgery of doctor appointments, car maintenance, parent-teacher conferences, paying bills, and cooking dinner every night is keeping me from doing the "really important work." But this feeling reflects a faulty perception based on the modern notion that the public and the private, the professional and the domestic, are separate spheres. "Social reproduction"—the care of self and others that is required for society to function—goes largely unpaid or at best undervalued, because the market assumes that people will do it on their own, out of the goodness of their hearts, as a literal labor of love. This structure makes many of us feel unvalued, that we can never demonstrate love for our loved ones enough. Understanding these structures of undervalue and vulnerability is the first step toward liberation.

Some activist circles perpetuate the idea that social reproduction is less valuable than the grander actions out there in the external world. But care of self, care of others, and care of what we think of as "the small," the granular, the micro, is *all*. That's the stuff of personal and collective resilience, the stuff we are fighting for, and it's the most potent way to resist structural oppression. Interior work is not just *useful* for sustaining the external work; it is the *greatest form of resistance to* the unjust structures we seek to change. If the *effort* to thrive undermines our *ability* to thrive, we are serving our own oppression.

Approach the granular scale of your interior life with the seriousness and focus you dedicate to your commitment to a just planet. Ensure that your interior and exterior work are synergistic, breathing life into each other.

UPPING YOUR EMOTIONAL RESILIENCE MEANS ▶▶

Taking Care of Yourself So You Can Stay with the Work

☐ Examine your barriers. What assumptions about being "in the fight" are stifling your ability to participate?

☐ Remember you are part of a collective. The work won't stop if you take a nap.

- ☐ Reject the martyr complex: no one person will win, or lose, this battle on their own.

- ☐ Remember that the frontlines are your very own body.

- ☐ Reject binaries between self-care and care of others, selfishness and selflessness, interior and exterior work.

Conclusion

Feed What You Want to Grow

Our imagination is a resource we should fiercely protect. Our attention, too, is precious. We frivolously give it away and let others take it at a great cost to ourselves, those around us, and the Earth. Our attention is fertilizer, sunlight, and water, encouraging whatever it focuses on to thrive. Literary critic Thomas Princen advises us how we can nourish both ourselves and our communities:

> We can "attend" to the drop-down menu's "Latest Headlines" and inhabit a world of bombings and coups, gadgets and personalities. Or we can attend to the people around us, to the institutions that shape us and that, to an extent, we shape, to the river that flows constantly but with great variability, to the buds that open every spring. We can attend to the language we speak: asking ourselves, is this the language of reductionism, domination, extraction, separation, consumerism and hence part of the problem? Or is it the language of holism, pluralism, co-creation, restraint, tending, ecosystems, inclusion, connection, place, community and hence a move in the right direction? In this sense, then, each of us can choose our world and, partly, create it.

"Here's your problem—it looks like you're paying attention to what's going on."

FIGURE 5. "Paying Attention," by Pia Guerra, *New Yorker.* Reprinted by permission from Conde Nast.

Train your thoughts and actions to be like sunlight and rain. Nurture the qualities, understandings, and attitudes that build resistance to climate change. Discipline your mind to choose and create the world you desire.

Paying attention to what is going on in the world today can cause dis-ease—emotional, physical, and existential (figure 5). But it is possible to stay calm and carry on amid unprecedented anxiety and global uncertainty. Focus your energy on the things that are going well. Amplify them. Hidden in the shadows cast by the extremists we see on social media are reasonable, well-intentioned people and enterprises making efforts and proposing ideas that need your time and attention: Men opposing sexualized violence. Engineers without borders. Artists inviting us to visualize better worlds. Emotional intelligence leaders influencing schools and businesses. Evangelicals for climate justice. Indigenous language revitalization immersion schools. Climate justice youth leadership conferences. Transformational resilience campaigns. The Just Transition and Movement Strategy collectives. The right-leaning millennials who formed the American Conservation Coalition. How can we be anything but heartened by the fact that so many people are profoundly concerned, not just about climate change, but about the social and ecological issues that spiral from it as well?

News and social media companies are not to blame for hawking stories that exploit our fear and survival instincts. But neither do we do have to buy what they're selling. Negative news not only doesn't reflect reality, it also blocks our ability to imagine a better world. Following Mister Rogers's mom's famous advice, always look for the people who are helping in a disaster, and choose to focus on them.

Resilience

People call my generation "Generation Z," as if we are the last generation. But we are not. We are refusing to be the last letter in the alphabet.

—JAMIE MARGOLIN, testifying before the House Foreign Affairs Committee, 2019

The tools I've explored in this book explain how to cultivate personal and collective resilience to create the best conditions for thriving in a climate-changed world. The term *resilience,* however, is itself problematic, as explained in the work of indigenous studies scholars. Some view "resilience" positively, saying it is what has kept indigenous cultures alive, despite many forms of colonial violence. These scholars reject the frame of victimization, shedding light instead on how people survive in the face of oppressive injustice. Recognizing "survivance" strategies, as indigenous scholar

Gerald Vizenor calls it, or "cultural continuance," the expression used by Potawatomi philosophy professor Kyle Powys Whyte, has been central to empowering indigenous identity.

Others, however, argue that the term paves the way for further injustice. If a community is "resilient," exploitation and damage are accepted as inevitable. Resilience can be used to justify leaving oppressive structures intact, since people are capable of learning to adjust to them.

Whyte and anticolonial activist-scholar Jaskiran Dhillon argue that the climate movement has appropriated indigenous forms of resilience. That is, those who have benefited from settler colonialism and are only now experiencing the loss and degradation of the lands they have settled on are extracting yet another resource from indigenous peoples—their strategies of "resilience." Dhillon asks, "whose interests are being served by attempts to extract and distill bits and pieces of Indigenous knowledge to work in the service of climate recovery?" Put another way, it is perfectly possible that white climate advocates may claim affinity with many aspects of indigenous knowledge—such as awareness of the interconnectedness of all life— without ever considering decolonization as a key to achieving their ideals.

Resilience must be advocated for in culturally sensitive ways, with acknowledgement that the crises of environmental change have been impacting indigenous peoples around the globe since the age of expansion, beginning as early as the fifteenth century. Climate change is not an impending *future* crisis. It is an extension of ongoing extinctions, destabilization, and rapid environmental transformation. We should resist crisis narratives for the reasons I have presented in this book, but also because they perpetuate the erasure of these legacies.

The term *resilience,* unlike *hope* or *optimism,* has the capacity for accepting negative feelings and legacies of ongoing crisis. As literary scholar Paul Outka puts it, "Resilience is a post-despair environmentalism, which isn't (at all) the same as optimism or thinking it's all going to work out fine—it's finding yourself still alive the day after, with some fight left. Resilience knows deeply how the best can be the enemy of the good, or even of the okay. Rather than trying to get it right, or get it right again, resilience is trying to keep it going here, there, anywhere, everywhere."

The wisdom to know the difference between problems we can control and those we cannot is essential for healing dis-ease. But navigating the gray area between these poles is complicated. While we cannot personally hold back sea-level rise, neither can we ignore it. Paying attention to that which we *cannot* control without losing the will to work on things we *can* takes an incredible amount of resilience. So too does letting go of things we have no command over—learning how to sit with your grief and sense of impotence, even while finding efficacy and healing in your power to act. Figuring out what you want to grow is a start.

Resistance

I'm with you when you say that climate change is the most important issue facing humankind. . . . . But, when I hear folks say—and I have heard it— that the environmental movement is the first in history to stare down an existential threat, I have to get off the train. . . . For 400 years and counting, the United States itself has been an existential threat for Black people. . . . I want you to understand how overwhelming, how insurmountable [slavery] must have felt. I want you to understand that there was no end in sight. It felt futile for them too. Then, as now, there were calls to slow down. To settle for incremental remedies for an untenable situation. They, too, trembled for every baby born into that world. . . . You don't fight something like that because you think you will win. You fight it because you have to.

—MARY ANNAÏSE HEGLAR, "Climate Change Ain't the First Existential Threat," 2019

Resilience is not just a way to take care of our own selves and hold on to secure and familiar routines as the world collapses. Political theorist Mark Neocleous warns that "resilience training" can actually support the status quo, that "resilience is by definition *against resistance*. Resilience wants acquiescence, not resistance." We must be on guard against such complacency. As climate disruption exacerbates existing structural oppression, such as racism, sexism, and economic injustice, we need to combat these structures in our daily lives. Thus the resilience I'm advocating is in fact *bound up with* resistance.

What does resilient action look like? It could start with taking care of yourself so you can go on with your life. But ultimately it must become a mutually reinforcing dynamic between the personal and the collective, since part of our personal resilience results from the sense of community that comes from participating in civic life. Everything from voting for candidates we believe in to participating in a world-building culture shift—it all matters, even if we can't see the immediate effects.

. . .

As an alternative to despair and hope, as a habit of mind to counter the distractions and addictions of modern life, as a way to draw on our own passions and strengths to effect change in the world, as a way to witness suffering without becoming incapacitated, as a way to enhance community and personal well-being while also challenging structures of oppression, and as an analysis of power that also focuses on daily practices, resilience helps us hold competing truths, see the world in a new way, and find the strength to keep getting up in the morning.

This critical moment in history calls us to imagine something larger than ourselves. When I was young, I didn't have a pressing cause to defend, and I felt I made no difference to anyone or anything beyond my immediate family. The same cannot be said of your generation. You have something to fight for. As a poet of our time, Mary Oliver, asks, "Tell me, what is it you plan to do with your one wild and precious life?" My hope for you is that this call to action gives you a sense of purpose, a reason to engage in the world, and an unwavering awareness of your power to build the world you deserve.

Acknowledgments

I want to acknowledge the support that my friends, colleagues, and family gave me over the course of producing this book. Thanks to Ronnie Chausse, Janet Fiskio, John Foran, Summer Gray, Jane S. Jaquette, Michael Lowenthal, Lenya Quinn-Davidson, James Ray, Abigail Reyes, Betsy Rosenbaum, the UC-CSU Transformative Climate and Sustainability Action and Education Knowledge-Action-Network, Humboldt State University (HSU) for the sabbatical time to write, and supportive colleagues Janelle Adsit, Stacy Becker, Renée Byrd, Loren Collins, Laura Johnson, Morgan King, John Meyer, and Rosemary Sherriff.

A handful of students who have powerfully shaped these ideas deserve special recognition, although this list is sure to be incomplete: Madi Whaley, Drew Andrew, Ty McCarthy, Nich Graham, Ryan Sandejas, Cela Wexler, Sam Stone, Anais Southard, Shanti Belaustegui Pockell, Emily Owen, Ben Nguyen, Shayan Farzadpour, Colin Mateer, Jack Davis, Sam Weeks, Carlrey Delcastillo, Chaya Zivolich, Emma Stokes, Ivan de Soto, Shiloh Green, Molly Gilmore, Andrew Jacobsen, and Ava Iorizzo.

Thank you to hosts who have invited me to talk about this topic: Mike Dronkers and HSU's MarComm Department; Giovanna di Chiro, Steve Hopkins, and Mark Wallace at Swarthmore College; the Interdisciplinary Anthropocene Studies Cluster at the University of Washington; the Rachel Carson Center; and the Walter H. Capps Center and the EJ/CJ Digital Hub at the Orfalea Center for Global and International Studies at UC Santa Barbara.

I am indebted to collaborators in this work who have helped open these conversations: Jennifer Atkinson, Kyle Bladow, Krista Hiser, Laura Johnson, Elin Kelsey, Laura Schmidt, and Nicole Seymour. Jennifer Ladino deserves special recognition for being the person who told me I should research and write about this stuff, and who provided fabulous revisions to earlier drafts. It is a great act of generosity to dive so deeply into another person's head and heart as she has with me.

I couldn't do anything without coffee shops providing space, internet, and the juice required for writing: Mosgo's, the Beachcomber, and Ramone's. The community of the Association for the Study of Literature and Environment is everywhere here—my peeps.

The interviewees of *Big Planet, Big Feels* enriched these ideas beyond Madi Whaley's and my hopes for that podcast project.

I feel fortunate to have worked with the team at University of California Press, led by tireless advocate and brilliant editor Stacy Eisenstark, who always "got" every little detail of what I was trying to do here and brought my ideas to life, literally. Tamie Parker Song, Dore Brown, Anne Canright, and Katryce Lassle were the last links in the chain, helping me express those ideas clearly and effectively.

The biggest gratitude of all goes to my students and other members of the climate generation who are demanding that we work to create a future we can all thrive in.

Notes

Introduction

p. 1 **This "vision-change-action" exercise:** I first learned about this exercise as a participant in four 2017 workshops with a group of University of California and Cal State University faculty called the Transformative Climate and Sustainability Education and Action Knowledge Action Network, or KAN. Funded by the UC Office of the President's Carbon Neutrality Initiative, this series of workshops was designed to build a network of scholar-activists to share teaching strategies for climate justice across California's public higher-education schools. The workshops were facilitated by Abigail Reyes, who used the vision-change-action model, and introduced us to the work of adrienne maree brown. brown's book *Emergent Strategy* came out in print shortly after our workshops.

p. 2 **snowflakes:** In polarized public discourse, college students are often represented as elite, emotionally weak "snowflakes." This representation gained ground especially after the 2016 presidential election, when right-wing news outlets showed condescending images of college students weeping over the outcome. The notion that college students are emotionally ill-equipped emerged even before that. In September 2015, an article by George Lukianoff and Jonathan Haidt in the *Atlantic,* "The Coddling of the American Mind," sparked a debate about whether colleges foster a "cult of fragility." I agree with and draw on Lukianoff and Haidt's 2018 book by the same title to describe the climate generation, but like them, I blame iGen members neither for the challenging contexts they face nor for the incomplete training they are getting to address climate disruption. Jennifer Atkinson responds directly to the right-wing attack on students who experience *climate* anguish in particular, in her 2018 piece "Addressing Climate Grief Makes You a Badass, Not a Snowflake" in *High Country News*.

p. 2 **We live in the Anthropocene:** The concept of the Anthropocene was popularized by atmospheric scientist Paul Crutzen in 2000, though it had been discussed before. While scientists still debate whether we are actually in the Anthropocene, the concept has permeated other disciplines. The ethical, moral, and existential implications of the possibility that humans have changed the chemical, biological, and geological nature of the planet have become topics of great interest to a wide range of thinkers, activists, and artists. See, for example, Clark, *Ecocriticism on the Edge;* Ghosh, *The Great Derangement;* and Scranton, *Learning to Die in the Anthropocene* and *We're Doomed. Now What?*

p. 3 **This is "the climate generation":** Data shows considerable consensus within Generation Z regarding the fact and urgency of climate change; indeed, the concern cuts across political divides, as Funk and Hefferon note ("Many Republican Millennials Differ"). Podcasts also offer insight, for example the 2017 Climate One episode "Inheriting Climate Change: What Will Boomers Leave Behind?" and the 2018 Global GoalsCast podcast episode "The Next Generations: We Can't Save the World without Them." Dunlap and Cohen's *Coming of Age at the End of Nature* is a great resource for thinking about some of these same questions in terms of Millennials.

p. 3 **you will owe on average $40,000:** See, for example, Institute for College Access and Success, "Student Debt and the Class of 2018," https://ticas.org/affordability-2/student-aid/student-debt-student-aid/student-debt-and-the-class-of-2018.

p. 3 **less well off than your parents:** Roos, "Are Millennials Really the First Generation to Do Worse than Their Parents?"

p. 4 **more lonely, suicidal, and depressed:** For more on the mental health picture of the climate generation, see Chowdry, "Research Links Heavy Facebook and Social Media Usage to Depression"; Doppelt, *Transformational Resilience;* Ducharme, "More than 90% of Generation Z Is Stressed Out"; Lukianoff and Haidt, "The Coddling of the American Mind"; Petersen, "How Millennials Became the Burnout Generation"; Reilly, "Record Numbers of College Students"; Rosenberg, "How We Can Help Students Survive an Age of Anxiety"; and Twenge, *iGen*.

p. 4 **By the time you reach college 65 to 85 percent of you:** Davidson, "Trauma-Informed Practices for Postsecondary Education: A Guide," p. 5.

p. 4 **least likely to vote:** "Generation Z Stressed about Issues in the News but Least Likely to Vote."

p. 4 **In 2018, the United Nations Intergovernmental Panel on Climate Change:** The IPCC is an intergovernmental body of the United Nations that collates and analyzes all scientific research on climate change. A "Summary for Policymakers" in the reports is often used to guide both international and regional policy. The 2018 report predicted faster increases in temperatures than had been previously reported. It also outlined predictions for social conflict and exacerbated inequalities.

p. 5 **"slow violence":** In *Slow Violence and the Environmentalism of the Poor,* Rob Nixon characterizes the lengthy suffering caused by environmental events that do not capture news media attention for long as "slow violence." For example, we think of a hurricane as a discrete event, but the conditions that exacerbate its effects, which continue well before and after contact, are rarely covered by the media; as a consequence, the public does not recognize hurricanes as part of a long-term, slowly unfolding, protracted crisis. The temporal stretch of climate change fails to capture the imagination or make it into news stories, Nixon suggests, precisely because its effects are felt over time.

p. 6 **change is coming, and the climate generation is driving it:** Aylin Woodward captures this momentum in an article titled "Millennials and Gen Z Are Finally Gaining Ground in the Climate Battle—Here Are the Signs They're Winning."

p. 7 **Conservatives in Gen Z care about climate change:** Funk and Hefferon, "Many Republican Millennials Differ with Older Members on Climate Change and Energy Issues"; Cohen, "The Age Gap in Environmental Politics"; Teirstein, "Seeing Red on Climate."

p. 7   At least **70 percent of your generation:** Reinhardt, "Global Warming Age Gap."

p. 8   ***imagine* the future we hope to live in":** Eaton, "Navigating Anger, Fear, Grief, and Despair," p. 48.

p. 9   **(80 percent of which are framed in negative terms):** Author interview with Elin Kelsey, which was made into the podcast episode "Elin Kelsey on Contagious Hope, Children's Books, and Solutions."

p. 9   **well situated to help fill this gap:** This growing subfield of environmental studies centers humanistic disciplines on environmental issues (e.g., environmental literary studies, environmental history, environmental philosophy), with a focus on actionability. Among those environmental humanists rethinking pedagogy based on the climate crisis and urging educators to consider students' emotions more actively are SueEllen Campbell, who led a June 23, 2015, pre-conference workshop on this topic at the Association for the Study of Literature and Environment (www.aslebiennialconference.com/teaching-climate-change .html), and Stephen Siperstein, Shane Hall, and Stephanie Lemanager, who co-edited an essential book for this subfield, *Teaching Climate Change in the Humanities.* Many environmental humanists are the first to be thinking about affect and the environment. Pertinent works include Donna Haraway's *Staying with the Trouble; Affective Ecocriticism,* edited by Kyle Bladow and Jennifer Ladino; Alexa Meik von Mossner's *Affective Ecologies;* Nicole Seymour's *Bad Environmentalism;* and Heather Houser's *Ecosickness in Contemporary U.S. Fiction.* Contemplative environmental practices are another focus, as explored for example in a 2017 book by the Evergreen collaborators Marie Eaton, Holly Hughes, and Jean MacGregor, *Contemplative Approaches to Sustainability in Higher Education,* or by Paul Wapner both in journal articles and in an annual workshop for educators and activists, "Contemplative Environmental Practice." The turn toward social sciences and humanities in environmental studies has come alongside the growing recognition that solutions to environmental problems will not be merely technological or scientific, but cultural, ethical, and social as well. The humanities and arts have an equally important role to play in the survival of the planet.

p. 9   **there are already many resources available:** For example, Boyd and Mitchell, *Beautiful Trouble;* Brecher, *Against Doom;* and Kalmus, *Being the Change.*

p. 10   **"the age of overwhelm":** Laura van Dernoot Lipsky's 2018 book by this title (subtitled "Strategies for the Long Haul") articulates the unique qualities of contemporary life that result in a widespread sense of "overwhelm"—a useful term for grasping what it feels like to exist in this political-historical moment. The term "climate overwhelm" is even more specifically about what living in the Anthropocene feels like for many.

p. 10   **"adaptive" strategies as opposed to "technical" solutions:** As the editors of *Contemplative Approaches to Sustainability in Higher Education* write, "Addressing technical problems is a matter of marshaling enough resources and expertise. In contrast, adaptive challenges are problems whose solutions are elusive because the problems themselves are so large, so highly complex, and so continuously evolving" (p. 5). Jem Bendell's groundbreaking 2018 essay "Deep Adaptation: A Map for Navigating Climate Tragedy" further refined this distinction by looking at the existential and spiritual implications of social collapse brought on by unfolding ecological collapse. *A Field Guide to Climate Anxiety* is squarely focused on these kinds of "deep adaptive" rather than technical solutions. Yet this distinction also opens the door for more robust engagement with social movements and grassroots strategizing, for it interrogates how "problems" are defined in the first place.

p. 10   **the feelings I explore in this book, such as climate anxiety, environmental trauma, and eco-grief:** A proliferation of essays on this topic reflects how much people are seeking to understand it. This list is not comprehensive, but shows the mounting attention in a range of media to climate anxiety and eco-grief: "It's Time to Talk about Ecological Grief" (UnDark), "'Climate Grief': The Growing Emotional Toll of Climate Change" (NBC News), "The Existential Dread of Climate Change" (Psychology Today), "Growing 'Ecological Grief' Is the Mental Health Cost of Climate Change" (CBC Radio), "Climate Change's Toll on Mental Health" (American Psychological Association), "Is The Changing Climate Giving You Anxiety? You're Not Alone" (KRCC),

Climate Change Is Causing Ecoanxiety and Damaging Your Health—
What Can We Do?" (Metro.co.uk), "Hope and Mourning in the Anthropocene: Understanding Ecological Grief" (The Conversation). Ashlee
Cunsolo and Neville Ellis's 2018 article "Ecological Grief as a Mental
Health Response to Climate Change–Related Loss" prompted much of
this newer interest, although people have been studying grief, mourning, anxiety, and despair as affects relating to the environment for some
time. See, for example, Fritze et al., "Hope, Despair, and Transformation: Climate Change and the Promotion of Mental Health and Wellbeing"; and Kevorkian, "Environmental Grief." Glenn Albrecht's notion of
"solastalgia" is a foundational term combining a sense of homesickness
with grief specifically related to environmental degradation. Also see
his 2019 book *Earth Emotions: New Words for a New World,* containing
more general work on "psychoterratica."

p. 11 **environmental justice:** In contrast to "mainstream environmentalism," environmental justice seeks to show the ways in which environmental degradation is linked to issues of social injustice. Drawing on
civil rights history and practices (Robert Bullard's 1990 book *Dumping
in Dixie: Race, Class, and Environmental Quality* is an informative history), environmental justice is both a scholarly and political-activist
movement and has helped shape the climate movement into the climate justice movement.

1. Get Schooled on the Role of Emotions in Climate Justice Work

p. 18 **Madi:** The stories about my students in this book are all true, but I have
changed their names for anonymity. Madi Whaley is the only excep-
tion; this is her real name. Madi courageously published the piece I
quote here and gave me permission to use her story.

p. 18 **Civil rights attorney and climate activist David Buckel:** Correal,
"What Drove a Man to Set Himself on Fire in Brooklyn?"

p. 18 **Photographer Chris Jordan:** Ahearn, "A Seattle Filmmaker Confronts
His Grief over a World That's Changing."

p. 19 **A viral YouTube video:** "Boy Crying Over Environment and People
Who Throw Trash on the Ground," YouTube, www.youtube.com
/watch?v=7x3Qh3_27Xk.

p. 19 **"influence what we choose to eat or how we get to work":** Kelsey, "Introduction" p. 5.

p. 19 **pre-traumatic stress disorder, solastalgia, and eco-grief:** On these and related concepts, see Albrecht, *Earth Emotions;* Albrecht et al., "Solastalgia"; Clark, *Ecocriticism on the Edge;* Cunsolo and Ellis, "Ecological Grief as a Mental Health Response"; Davenport, *Emotional Resiliency in the Era of Climate Change;* Doppelt, *Transformational Resilience;* Eisenstein, *Climate;* Green, "The Existential Dread of Climate Change"; Houser *Ecosickness in Contemporary U.S. Fiction* and *Environmental Culture of the Infowhelm;* Kelsey, "Introduction"; Kevorkian, "Environmental Grief"; Ladino, *Memorials Matter;* Lipsky, *The Age of Overwhelm;* Maniates, "Teaching for Turbulence"; Morton, "Don't Just Do Something, Sit There!," *The Ecological Thought,* and *Hyperobjects;* Scher, "'Climate Grief': The Growing Emotional Toll of Climate Change"; Solnit, *Hope in the Dark,* "'Hope Is an Embrace of the Unknown,'" *The Mother of All Questions,* and *Paradise Built in Hell;* Spark, "Mourning the Ghost"; Weston, *Mobilizing the Green Imagination;* and Whaley, "The Beautiful Environmentalist." Some of these terms are associated with particular thinkers: Albrecht coined the term *solastalgia,* Kevorkian copyrighted *ecological grief,* and pre-traumatic stress disorder was defined by Greta van Susteren, co-founder of the Climate Psychology Alliance. Allison Ford and Kari Norgaard's "From Denial to Resistance: How Emotions and Culture Shape Our Responses to Climate Change" and the online Climate Psychology Alliance attest to the growing attention to the relationship between emotions, mental health, and climate change.

p. 19 **a feeling of dread about the future:** On the indirect mental health effects of climate change, see Sliwa, "Climate Change's Toll on Mental Health." New mental health concerns are also arising due to responses to the challenges of coping with climate change—what some call "knock-on," or secondary, effects, ranging from adaptation (such as improved health care services as a way to adapt to increased physical and mental health problems) to prevention (such as improved infrastructure and environmental regulations to avoid problems in the first place). See Doppelt's *Transformational Resilience* for further discussion of the ripple effects of climate change on social and emotional support structures.

p. 19 **Climate change disasters have increased rates of suicide, depression, and anxiety:** The following statistics are from Clayton et al. (eds.), *Mental Health and Our Changing Climate: Impacts, Implications, and Guidance;* and Burke et al., "Higher Temperatures Increase Suicide Rates in the United States and Mexico."

p. 20 **vicarious trauma, compassion fatigue, or apocalypse fatigue:** Leslie Davenport (*Emotional Resiliency,* pp. 112–13) defines vicarious trauma as "the cumulative emotional impact that results from empathic engagement with traumatic experiences. It is also known as compassion fatigue and secondary traumatic stress." Apocalypse fatigue as a concept has gained more attention thanks to Per Espen Stoknes's TED Talk "How to Transform Apocalypse Fatigue into Action on Global Warming."

p. 20 **gives names to the feelings people have:** See Albrecht et al., "Solastalgia: The Distress Caused by Environmental Change"; and Albrecht, *Earth Emotions: New Words for a New World.*

p. 21 **"anticipatory grief":** Spark, "Mourning the Ghost," p. 27.

p. 21 **weakened immune system:** Swaminathan et al., "Will Global Climate Change Alter Fundamental Human Immune Reactivity: Implications for Child Health?"

p. 22 **Congresswoman Alexandria Octavio-Cortez:** See Irfan, "We Need to Talk about the Ethics of Having Children in a Warming World."

p. 22 **The new #BirthStrike community:** The quotation is from https://birthstrike.tumblr.com. See also Hunt, "BirthStrikers"; and Rainey, "For Some Millennials, Climate Change Clock Ticks Louder Than Biological One."

p. 22 **queer ecological studies:** See Seymour, *Strange Natures;* di Chiro, "Polluted Politics?"; and Sandilands, "Queer Ecology."

p. 23 **Current indigenous scholarship:** See, for example, Dhillon, "Indigenous Resurgence, Decolonization, and Movements for Environmental Justice."

p. 25 **Western science:** Science and technology studies and feminist science studies address how culture shapes scientific inquiry, especially with attention to power and gender. Some exemplary works are Peter Bowler's *A Norton History of Environmental Sciences,* Donna Haraway's

Primate Visions, Sandra Harding's *Whose Science? Whose Knowledge?,* and Bruno Latour's *We Have Never Been Modern.*

p. 25 **have had cause to be suspicious of Western science:** See, for example, Corburn, *Street Science;* Di Chiro, "Environmental Justice"; and Black et. al., *A Line in the Tar Sands.*

p. 25 **Historically, Western science has also been misused:** For understanding the role of Western research in exploiting or erasing indigenous knowledges, see Linda Tuhiwai Smith's *Decolonizing Methodologies.* The role of Western science in proving inferiority is explored in Ray, *The Ecological Other.*

p. 25 **traditional ecological knowledge:** Here, from the robust subfield of indigenous studies, I draw in particular on Whyte, "Indigenous Science (Fiction) for the Anthropocene" and "Is It Colonial Déjà Vu?"; and Kimmerer, *Braiding Sweetgrass.* Jason Corburn's *Street Science* also argues that lay knowledge ought to help shape scientific knowledge.

p. 27 **climate change communication:** In the scholarly field of climate change communication, two key texts are Stoknes, *What We Think about When We Try Not to Think about Global Warming,* and Marshall, *Don't Even Think about It.* The Yale Center for Climate Communication is a central hub of research and action in this area, known for its biannual study *Climate Change in the American Mind.*

p. 27 **the "marathon" (as opposed to "sprint") of social change:** The work of changing culture and transformative justice takes time, as environmental justice leaders such as Robert Bullard, Grace Lee Boggs, and others have argued (also see Brian McDermott's film *Marathon for Justice*). They theorize the slow pace of social change as a way to remind activists that results of their labors are not always evident or immediate. Paul Loeb's collection of essays by activist leaders and artists, *The Impossible Will Take a Little While,* unpacks this notion of why social change is slow. These ideas work in tandem with those who are thinking about the role of contemplative practice in environmental work, such as Marisol Cortez, who criticizes the activist community for succumbing to the unsustainable work pace of capitalist productivism; humanities scholars like Christof Mauch and Timothy Morton, who promote the importance of thought, theory, and critique to any worthwhile "action"; and the many emotional intelligence advocates and risk psychologists I mention in this book, who ask that we both pay closer attention to and learn how to better manage our emotions in response to perceived risks.

p. 28 **Insights from trauma studies and social movements:** A key contribution of social movement theory for the climate movement is its awareness of the role of trauma in shaping the terms and goals of social change. I'm drawing here on Laura van Dernoot Lipsky's *Trauma Stewardship* and David Treleaven's *Trauma-Sensitive Mindfulness*. In this section, I'm only briefly exploring the question, how might the traumatic effects of climate change—"climate trauma"—shape or mobilize the climate generation? Inversely, and perhaps even more importantly, how might racialized, sexualized, or historical trauma shape people's relationship to the climate generation?

2. Cultivate Climate Wisdom

p. 31 **affect theory:** See generally Sara Ahmed, *The Promise of Happiness,* and Lauren Berlant, "Cruel Optimism." For examples of how affect theory is being taken up in the environmental humanities, see Bladow and Ladino (eds.), *Affective Ecocriticism;* von Mossner, *Affective Ecologies;* Houser, *Ecosickness in Contemporary U.S. Fiction;* and Seymour, *Strange Natures* and *Bad Environmentalism.*

p. 31 **secular mindfulness:** Although this notion draws on aspects of Buddhist, especially Zen, mindfulness, one need not subscribe to a religion to practice mindfulness. Whole subspecies of mindfulness are emerging in workspaces, educational settings, and therapies. My own personal resources include Judson Brewer's *The Craving Mind;* Jon Kabat-Zinn's *Mindfulness for All* and *Wherever You Go, There You are;* Shunryu Suzuki's *Zen Mind, Beginner's Mind;* and Tara Brach's *Radical Acceptance* and *True Refuge* (see also www.tarabrach.com).

p. 31 **eco-psychology:** Some of the most important research on emotions and climate change is coming from eco-psychology (see Davenport, *Emotional Resiliency in an Era of Climate Change*), environmental sociology (see Norgaard, *Living in Denial: Emotions, Climate Change, and Everyday Life*), the emerging science of emotion (see the work of Marc

Brakett's Yale Center for Emotional Intelligence [ei.yale.edu] and the field of neuroaffective sciences), and mindfulness (see teachings of the secular teacher and therapist Tara Brach, mindfulness expert David Treleaven [e.g., *Trauma-Sensitive Mindfulness*], and cognitive psychologist Judson Brewer), arguing that we need to take emotions much more seriously in climate discussions. Increasingly, the environmental aspects of mental health are shaping discussions in the fields of secular mindfulness, contemplative practices, and trauma studies, such as can be found in Eaton, Hughes, and MacGregor (eds.), *Contemplative Approaches to Sustainability in Higher Education,* for example.

p. 32 **scientists suffer some of the highest rates of eco-despair:** See Clayton, "Mental Health Risk and Resilience among Climate Scientists."

p. 33 **the social sciences are proving:** David Ropeik, "It's Time for Climate Change Communicators to Listen to Social Science."

p. 33 **emotional intelligence (EI):** Research from the Yale Center for Emotional Intelligence (ci.yale.edu), Lukianoff and Haidt's book *The Coddling of the American Mind,* UC Berkeley's Greater Good Science Center (greatergood.berkeley.edu), and scholars focusing on social-emotional learning (SEL) have informed my thinking on this. In the field of child development, SEL is the pedagogy *du jour,* and the Yale Center for Emotional Intelligence focuses these insights on teens; less attention meanwhile has been paid to the importance of emotional intelligence in the educational setting of college or among college-aged adults. Meanwhile, information about the mental health profile of iGen is growing, which ought to inform the strategies of those of us who work closely with iGen. Almost all campuses hire counseling and psychological services staff, but that's not the same thing as acknowledging that students are emotional beings *in college classes* or how important emotion is to fostering conditions for learning. On the contrary, we expect students not to be emotional in class, much to the detriment of the very reason we are there: *to facilitate student learning.* Against all the research that proves that people learn better when they are emotionally invested, both in the content and in relationships in the classroom, we expect students to learn eagerly, demonstrate their learning as individuals, and retain what they learn forever—all without considering how they feel when they come into and leave class. For a source on bringing emotional intelligence into the college classroom, see Sarah Cavanaugh's *The Spark of Learning*. For more on centering emotion in the classroom as anti-oppressive pedagogy, see Duncan-Andrade, "Note to Educators"; hooks, *Teaching Community;* Orr, "The Uses of Mindfulness in Anti-Oppressive Pedagogies"; Thompson, *Teaching with Tenderness;* and Paris and Alim (eds.), *Culturally Sustaining Pedagogies.* A few scholars are working on how to center students' emotions in environmental classrooms. See my "Coming of Age at the End of the World"; Barker and Franklin, "Social and Emotional Learning for a Challenging Century" (source of quotation in text, p. 96); and Hufnagel, "Students' Emotional Connections to Climate Change." That said, there is also robust criticism of EI, especially in terms of how it has been co-opted by corporate culture as a strategy to leverage more productivity from employees.

p. 35 **who gets to determine which facts to pursue:** For a discussion of the contextualization, both political and social, of science, see, for example, Bowler, *Norton History of the Environmental Sciences.*

p. 35 **"infowhelm":** Houser, *Environmental Culture of the Infowhelm.*

p. 35 **bombard us with the specter of apocalypse:** Exemplary of this mode is David Wallace-Wells's *This Uninhabitable Earth,* for example.

p. 35 **"information deficit":** Norgaard's *Living in Denial* demonstrates that people are not always rational creatures, able to make decisions about how to act, vote, and invest based on data or information. For a long time, climate advocates were told to provide "more data" to prove the forecasts and eliminate uncertainty. Sociologists and psychologists are now proving that offering more data is not an effective way to mobilize concern; on the contrary, more information can inhibit action.

p. 36 **"When you don't trust anyone talking about climate change":** Quoted in Teirstein, "Seeing Red on Climate."

p. 37 **Denial and dissonance:** Climate change is uniquely difficult to get our emotional heads around. Eco-psychologists argue that this is because of "cognitive dissonance," whereby the discouraging stories about climate change rub against our positive desires for our daily lives. In *Memorials Matter,* literary critic Jennifer Ladino proposes the companion concept of "affective dissonance," which describes our conflicting emotional responses to climate change—for example, between our desire for life to stay the same and our horror in the knowledge of how it's already changed. Similarly, in *Engaging the Everyday*, John Meyer talks about the "resonance dilemma," the fact that many of us don't associate our daily experiences with the abstract, global phenomena of climate change that are coming down the road.

p. 39 **Is happiness getting in the way of our happiness?** In my critique of happiness I'm drawing on Ahmed, *The Promise of Happiness*; Berlant, "Cruel Optimism"; and the Harvard Study of Adult Development (see Mineo, "Harvard Study"). For more American context, see Florida, "The Unhappy States of America."

p. 40 **avoidance behaviors and emotional dysregulation:** Attention to negative or shameful feelings is necessary to achieve climate wisdom. This insight combines ideas about dysregulation in mindfulness psychology from Jon Kabat-Zinn, David Treleaven, Judson Brewer, and Tara Brach with concepts about climate psychology from writers like Roy Scranton, Leslie Davenport, and Per Espen Stoknes. Affect theorists like Sara Ahmed also contend that negative feelings have an important role to play in moving toward justice.

p. 41 **"The organizational culture of social justice nonprofits":** Cortez, "The Praxis of Deceleration," p. 78. Although I am drawing on Marisol Cortez's use of the term *productivism* here, it is also supported by scholarship in disability studies, which shows that disability is in fact created, not just excluded, by productivist imaginary. Also informative is the emerging notion of "slow scholarship"; see Berg and Seeber, *The Slow Professor*; and Mountz et al., "For Slow Scholarship." Cortez argues for "deceleration" to resist the burnout logic of both activism and capitalism.

p. 43 **"paying attention in a particular way":** Kabat-Zinn, *Wherever You Go, There You Are*, p. 4.

p. 43 **"By virtue of paying close, nonjudgmental attention":** Treleaven, *Trauma-Sensitive Mindfulness*, p. 36.

p. 47 **RAIN:** Tara Brach is a meditation teacher and psychotherapist bringing Western psychology together with Eastern spiritual practices, and the author of *Radical Acceptance* and *True Refuge*. She records her meditations and lectures and makes them available on her website and in a podcast. RAIN is her main psychotherapy/meditative tool; see www .tarabrach.com.

p. 47 **In Buddhism, self-regulation is built into the Eightfold Path:** Treleaven, *Trauma-Sensitive Mindfulness*, pp. 75–79.

p. 49 **small steps toward which we can "nudge" ourselves:** I am drawing on Judson Brewer's *The Craving Mind* as well as the *TED Radio Hour* episode "Nudge" (www.npr.org/programs/ted-radio-hour/483080945 /nudge). The concept suggests that behavioral change is hard, but people are more likely to start if they only have to do small things.

p. 50 **A mindfulness worksheet:** You can download a mindfulness worksheet from https://mindfulnessexercises.com/free-mindfulness-worksheets. Mindfulness apps, such as Headspace and Insight Timer, are also helpful, especially for guided or sound-related meditations.

3. Claim Your Calling and Scale Your Action

p. 52 **Ron Finley, the self-described "gangsta gardener":** The details and quotations in this paragraph are from http://ronfinley.com.

p. 53 **"Each one of us must choose":** Jamail and Cecil, "Rethink Activism in the Face of Catastrophic Biological Collapse."

p. 54 **"has its own power":** Eaton, "Navigating Anger, Fear, Grief, and Despair," p. 42.

p. 55 **"Small is good, small is all":** brown, *Emergent Strategy*, pp. 41–42.

p. 55 **"collective amnesia":** Loeb, *Soul of a Citizen*, pp. 44–45.

p. 58 **"building the capacity of people to cope with the traumas":** Doppelt, *Transformational Resilience*, p. 67.

p. 59 **"no response to the climate crisis":** Ibid., p. 67.

p. 59 **"First, we need to believe":** Loeb, *Soul of a Citizen*, p. 5.

p. 59 **"If you expect":** Quoted in ibid., p. 123.

p. 60 **"It's always too soon to go home":** Solnit, "Acts of Hope."

p. 60 **"revolutionary change":** Zinn, "The Optimism of Uncertainty," pp. 85–86.

p. 60 **In the instrumentalist view:** My critique of the instrumentalist assumptions behind "problem-solving" and "action" draws on Nicole Seymour's *Bad Environmentalism;* the essays in Loeb's *The Impossible Will Take a Little While;* Rebecca Solnit's "Acts of Hope"; adrienne maree brown's *Emergent Strategy;* Timothy Morton's "Don't Just Do Something, Sit There!"; Scott and Paul Slovic's *Numbers and Nerves;* Janet Fiskio's "Building Paradise in the Classroom"; Ted Toadvine's "Six Myths of Interdisciplinarity"; and Michael Maniates's "Teaching for Turbulence."

p. 61 **"good-enough activists:"** Loeb, *Soul of a Citizen,* pp. 49, 50.

p. 61 **Literary theorist Nicole Seymour argues:** Seymour, *Bad Environmentalism,* p. 57.

p. 63 **Seymour lists many actions that the instrumentalist approach to environmentalism overlooks:** Ibid., p. 7.

p. 64 **"distant in time and space":** Stoknes, *What We Think about When We Try Not to Think about Global Warming,* p. 40.

p. 65 **In *ReGeneration:*** *ReGeneration* is a publication of the Movement Strategy Center in Oakland, California, which offers tools for doing community and personal resilience-building work, including the Transition Community Initiative and resources on leadership, movement building, and environmental justice organizing.

p. 67 **"ecological thought":** Morton, *The Ecological Thought,* pp. 2, 9.

p. 67 **"Reframing our world, our problems, and ourselves":** Ibid., p. 9.

p. 67 **"We never quite fathom":** Slovic, *Going Away to Think.* For more on "slow thinking," see Kahneman, *Thinking, Fast and Slow.*

p. 67 **"stories of slow hope":** Mauch, "Slow Hope."

p. 67 **"international outpouring of marine conservation success stories":** See Elin Kelsey's essay "Birth of Ocean Optimism."

p. 68 **"difficult to take time to be inclusive":** See *Dismantling Racism 2016 Workbook,* https://resourcegeneration.org/wp-content/uploads/2018/01/2016-dRworks-workbook.pdf, p. 29. Kyle Powys Whyte also makes this case in "Is It Colonial Déjà Vu?," in which he argues that the politics of urgency has made climate change advocacy largely unrelated to indigenous efforts to achieve justice (p. 34).

p. 68 **"Coercive conservation":** For more on conserving natural resources or wildlife at the cost of social justice, see Mark Dowie's "Coercive Conservation"; Mark Spence's *Dispossessing the Wilderness;* Richard Grove's *Green Imperialism;* and the work of myriad critical environmental historians, including but not limited to Carolyn Merchant (e.g., *The Death of Nature*), Diana K. Davis (e.g., *Resurrecting the Granary of Rome*), and Stephen Germic (e.g., *American Green*).

p. 68 **In the late nineteenth century:** The frontier thesis was proposed by Frederick Jackson Turner in 1893 (see Slotkin, *Gunfighter Nation*). The social anxieties around definitions of American identity that the thesis tapped into and exacerbated helped lay the foundation for the creation of national parks and wilderness areas, under the leadership of Theodore Roosevelt and Gifford Pinchot, but led to genocide of indigenous peoples, among many other social injustices. See Spence, *Dispossessing the Wilderness;* and Ray, *The Ecological Other.*

p. 69 **at least one and possibly over two million organizations:** Hawken, *Blessed Unrest,* p. 2.

p. 69 **"relational, adaptive, fractal":** brown, *Emergent Strategy,* p. 36.

p. 70 **"is taking shape in schoolrooms, farms, jungles":** Hawken, *Blessed Unrest,* p. 3.

p. 70 **"let that work be tangible":** brown, *Emergent Strategy,* p. 256.

p. 70 **"emphasizing the power of collective action":** Davenport, *Emotional Resiliency,* p. 136.

p. 70 **"solutions journalism":** Elin Kelsey is focused on "the contagious nature of hope" and works to "spread a global epidemic of solutions-focused environmental engagement." She created a platform for people to share stories about good things that are happening in the ocean at the Twitter hashtag #OceanOptimism. She has also worked with the Smithsonian Institution's research on solutions. From a fellowship with the Rachel Carson Center, she co-created an open-access collaborative site called *Beyond Doom and Gloom* (www.environmentandsociety.org /exhibitions/beyond-doom-and-gloom) as well as an online group-sourced syllabus, *Radical Hope.* In a podcast interview, "Elin Kelsey on Contagious Hope, Children's Books, and Solutions," she said that she doesn't worry about having hope, because she spends all of her time collecting stories of solutions. She therefore "breathes a different oxygen" than people who see only the bad news and who therefore operate from a position of despair. For examples of "solutions journalism," see www.thesolutionsjournal.com and www.yesmagazine.org.

p. 71 **"it is social capital"**: Davenport, *Emotional Resiliency*, p. 141. Davenport is citing the research of Daniel Aldrich, which showed that in communities of trust "fewer lives were lost" in times of disaster (see Aldrich and Meyer, "Social Capital and Community Resilience"). I draw on brown's *Emergent Strategy*, Davenport's *Emotional Resiliency*, Doppelt's *Transformational Resilience*, and Hawken's *Blessed Unrest* to explain why social capital is more effective than infrastructure in helping a community recover from a disaster.

p. 71 **"growing body of evidence"**: Doppelt, *Transformational Resilience*, p. 9.

p. 72 **"focus on critical connection"**: brown, *Emergent Strategy*, p. 42.

p. 72 **"people easily feel helpless"**: Stoknes, *What We Think about When We Try Not to Think about Global Warming*, p. 101.

p. 72 **A variety of other kinds of expertise and ways of thinking:** For more on the ideas discussed in this paragraph, such as "new materialism," "eco-phenomenology," "interbeing," see variously Kimmerer, *Braiding Sweetgrass*; Bennett, *Vibrant Matter*; Alaimo, *Bodily Natures*; Abrams, *The Spell of the Sensuous*; Toadvine, "Six Myths of Interdisciplinarity"; Hanh, *The Art of Living*; and Margulis, *Symbiotic Planet*.

p. 73 **"If you have come here to help me"**: Aboriginal activists group, Queensland, 1970s (for "authorship" of the phrase, see https://en.wikipedia.org/wiki/Lilla_Watson).

p. 73 **The concept of "pseudoinefficacy"**: For more on this concept, see www.arithmeticofcompassion.org/pseudoinefficacy.

p. 74 **"Suppose, as you see the child go under"**: Västfjäll, Slovic, and Mayorga, "Pseudoinefficacy and the Arithmetic of Compassion," p. 43.

p. 74 **"arithmetic of compassion"**: See an explanation of this concept at Slovic and Slovic, "The Arithmetic of Compassion." Herbert's poem can be found at Whetham, "James Johnson, 'The Arithmetic of Compassion': Rethinking the Politics of Photography."

p. 75 **"Perceived rather than actual efficacy"**: Ibid., 45.

p. 77 **"get really good at being intentional"**: brown, *Emergent Strategy*, p. 72.

p. 77 **"Critical Hope"**: www.enst490.blogspot.com.

p. 78 **toxic siting:** A pivotal study released in 1987 found that race, not class, was the leading factor in the siting of polluting industries. See the

Commission for Racial Justice, *Toxic Wastes and Race in the United States.*

p. 78 **"uprisings and resistance"**: brown, *Emergent Strategy*, p. 119.

4. Hack the Story

p. 80 **"more Americans can imagine the 'end of the world'"**: Norgaard, "Climate Change Is a Social Issue."

p. 80 **people will more often select—and remember—depressing stories**: Stafford, "Why Bad News Dominates the Headlines."

p. 81 **"Environmental solutions are emerging"**: Kelsey, "Introduction," p. 7.

p. 81 **"We live in capitalism"**: Ursula Le Guin, speech upon accepting the National Book Foundation's Medal for Distinguished Contribution to American Letters at the National Book Awards, November 19, 2014, available at www.sfcenter.ku.edu/LeGuin-NBA-Medalist-Speech.htm.

p. 82 **"Whether or not the world really is getting worse"**: Pinker, "The Media Exaggerates Negative News." Pinker calls this the "negativity effect." Negativity bias is the tendency of news media to exploit our fight-or-flight instincts with particular kinds of news stories. The scholarship I'm drawing on here includes but is not limited to Pinker's *Enlightenment Now,* Rob Nixon's *Slow Violence and the Environmentalism of the Poor,* Per Espen Stoknes's *What We Think about When We Try Not to Think about Global Warming,* and the chapter on the "fear instinct" in Rosling, *Factfulness.* Making better deliberate choices about what kinds of media you consume is one strategy for self-care.

p. 82 **our perceptions often don't match the probability of actual threats occurring**: Risk perception social psychology is helpful in understanding why people perceive some risks as threats but not others. What does it take to mobilize politicians and the public around a risk like climate change such that they invest in avoiding it? Here, I draw on the work of Paul Slovic, *The Feeling of Risk;* Ursula Heise, *Sense of Place and Sense of Planet;* Ulrich Beck, *Risk Society;* Per Espen Stoknes, *What We Think about When We Try Not to Think about Global Warming;* George Marshall, *Don't Even Think about It;* and Anthony Leiserowitz et al., *Climate Change in the American Mind.*

p. 83 **"You almost couldn't design a problem that is a worse fit"**: Anthony Leiserowitz, interviewed by Bill Moyers in "Ending the Silence on Climate Change," *Moyers & Company*, March 15, 2013, available at www .youtube.com/watch?v=mJx14zuSL10. See also episode 1 of *Climate Lab*, "Why Humans Are So Bad at Thinking about Climate Change," April 19, 2017, available at www.youtube.com/watch?v=DkZ7BJQupVA&vl=en.

p. 83 **"estimate the probability of an event"**: The quote is from Pinker, *Enlightenment Now*, p. 41, referencing A. Tversky and D. Kahneman, "Availability: A Heuristic for Judging Frequency and Probability," *Cognitive Psychology* 4 (1973): 207–32.

p. 83 **"Bad things can happen quickly"**: Pinker, *Enlightenment Now*, p. 41.

p. 83 **"the consequences of negative news are themselves negative"**: Ibid., p. 42.

p. 84 **other aspects of our lives . . . are on the upswing:** See Easterbrook, *It's Better Than It Looks*; Pinker, *Enlightenment Now*; and Rosling, *Factfulness*.

p. 84 **His conclusion is that the state of the world is improving:** Hans Rosling (*Factfulness*, p. 17) claims that factfulness can "make you feel more positive, less stressed, and more hopeful as you walk out of the circus tent and back into the world." In contrast to "apocalyptic storytelling," strategies of factfulness employ critical thinking to check our emotional responses to the "circus tent" of media stories. As Rosling points out, the fact that communication and news are more accessible than ever before in human history is a sign of human progress. Yet the media makes us constantly aware of the bad things that are happening everywhere, and so, ironically, although we are safer than ever, we feel more unease than ever. Simply put, factfulness is a critical thinking strategy to help us not worry more than we need to.

p. 85 **"factfulness worldview"**: Ibid., p. 13.

p. 85 **Rosling supplies a cheat sheet:** Ibid., p. 256.

p. 86 **terrible sixth extinction:** See, generally, Kolbert, *The Sixth Extinction*.

p. 87 **the jeremiad form:** Sacvan Bercovitch's *American Jeremiad* describes the unique features of the US litany of woes. Many environmentalist diatribes follow this format.

p. 89 **"ecological imagination"**: Norgaard, in "The Sociological Imagination in a Time of Climate Change," points out that while scientists have helped us gain an ecological imagination that shows us the relationship between humans and the earth, a sociological imagination would take this a step further to ask why climate change is happening, how we are being impacted, why we have failed to successfully respond so far, and how we might be able to effectively do so.

p. 90 **Your attention is like gold to them:** Listen to the TED Radio Hour episode "Attention, Please," May 26, 2018, www.npr.org/programs /ted-radio-hour/614007696/attention-please; and see the chapter "Less Distraction: More Intention" in Lipsky, *The Age of Overwhelm*. In *A Paradise Built in Hell,* Rebecca Solnit calls the co-optation of our attention through advertising "economic privatization," which, she says, "is impossible without the privatization of desire and imagination that tells us we are not each other's keeper" (p. 9). See also Haiven and Khasnabish, *The Radical Imagination,* for an example of how people have been convinced that the only power they have is as consumers.

p. 92 **Student Leadership Institute for Climate Resilience (SLICR):** The University of California's mandate to scale student leadership training throughout the ten campuses has also included interested California state universities, such as Humboldt State, where I work. Informed by Just Transitions, Movement Generation, and Movement Strategy techniques for social change, SLICR has been led by Abigail Reyes and Aryeh Shell. For more information, see http://communityresilience .uci.edu/uc-wide-trainers-training.

p. 92 **"Three Stories of Our Time":** Macy and Johnstone, *Active Hope,* 4–5; also at www.activehope.info/three-stories.html.

p. 93 **"When we find a good story and fully give ourselves to it":** Macy and Johnstone, *Active Hope,* p. 33.

p. 93 **Global Awareness class:** This assignment, to document social movements and solutions, was designed by Dr. Laura Johnson.

p. 94 **student food insecurity:** See, for example, Crutchfield and Maguire, "Study of Student Basic Needs," p. 8; and Williams, "Fighting Food Insecurity on College Campuses."

p. 94 **"involves the transition from a doomed economy":** www .activehope.info/great-turning.html.

p. 94 **62 percent of Americans were at least "somewhat worried":** Leiserowitz et al., *Climate Change in the American Mind.*

p. 94 **over 70 percent of Gen Zers agree:** Barenberg and Corzo, "Generation Z Is Not Afraid."

5. Be Less Right and More in Relation

p. 97 **ask with disgust:** I unpack the idea of "environmentalist disgust" in my book *The Ecological Other.* There I argue that disgust as an affect makes it easy for environmentalists to consciously or unconsciously disguise other forms of distaste in a morally righteous green package. This is particularly obvious in the current climate debate, where acting anti-environmentally (rolling coal, mocking Priuses, etc.) is less a statement on climate change than it is a rejection of liberal coastal elitism. Environmentalists may be disgusted by that behavior, only barely hiding class disdain behind green virtue. They may hypocritically associate groups of people with their trash or emissions, demonizing houseless people for depositing trash in the places they move through, for example, rather than recognizing the NIMBY privilege of having your trash moved from your curb to some other place—out of sight and out of mind, but still a major environmental problem. In *Bad Environmentalism: Irony and Irreverence in the Ecological Age,* Nicole Seymour says that climate skepticism should be read primarily as a rejection of environmentalist *class* elitism. Inversely, one reason that "the poor" are not seen as environmentalists is that they are not privileged, and so their choice to not consume does not carry the same moral valence as it does for someone who "sacrifices" luxuries in order to claim superiority. Disgust has often been about maintaining divisions between upper and lower classes, as Pierre Bordieu contends in *Distinction: A Social Critique of the Judgment of Taste.* Resisting elitist disgust has taken myriad forms in American cultural history, and being "disgusted" by non-environmentalist behavior misses the important socioeconomic messages in those behaviors. See Alexis Shotwell, *Against Purity: Living Ethically in Compromised Times,* for more arguments along these lines. The affect of disgust plays a problematic role in much environmental and anti-environmental identity negotiation.

p. 97 **we are shutting down constructive dialogue before it has even begun:** In what follows, I draw in part on episodes from the podcast *On*

Being: The Civil Conversations Project, such as "Relationship across Rupture," as exemplary of the kind of work people are trying to do to repair rifts.

p. 97 **highly educated scientists who use a specialized vocabulary:** "Climate change" itself is a term that can put people off. In "The Art of Arguing Science," Clare Follmann quotes linguist Uwe Poerksen on "plastic words": "When they first appear, they are fashionable and command attention; but they merge with the everyday and soon seem commonsense." They can mean anything and nothing at once. A plastic word "signals science. It silences" (p. 3). Although neither Follmann nor Poerksen explicitly discuss climate change, it fits the theory well. Its plasticity, its bigness, its foundation in claims to truth and scientific certainty, and its presentation by advocates as the greatest crisis ever all can undermine its engagement with social justice.

p. 98 **perhaps we should be asking what opportunities:** Thanks to Janet Fiskio for this and other insights in this chapter.

p. 99 **"Curious, Compassionate Conversation":** This assignment is based on Richard Rorty's insight that "speaking differently" is more important than "arguing well" in bringing about social change. I had been taught to teach students how to use rhetorical techniques and argumentation to persuade audiences, and so this assignment undid some of my own training. For the full, downloadable assignment, see "Curious, Compassionate Conversation" under my list of Assignments, Modules & Activities at UC-CSU NXTerra, www.nxterra.orfaleacenter .ucsb.edu/modules-assignments-and-activities-for-climate-change-emotions/.

p. 100 **the National Association of Evangelicals recognized climate change:** Yoder, "An Evangelical Leader Calls Young Christians to Save the Planet."

p. 100 **"Getting more people on board":** Hayhoe, "The Most Important Thing You Can Do to Fight Climate Change."

p. 100 **"engage the issue of climate change":** Dan Kahan, quoted in Stoknes, *What We Think about When We Try Not to Think about Global Warming,* p. 108.

p. 101 **"the places we have loved and lost":** Eisenstein, *Climate—A New Story,* pp. 49–50.

p. 101 **reframing it as a question of health:** Pellow, *What Is Critical Environmental Justice?* See also John Meyer, *Engaging the Everyday,* introduction; and Dunne, "Impact of Climate Change on Health Is 'the Major Threat of 21st Century.'" News reports about climate change increasingly mention asthma, emissions, and infectious disease, for example.

p. 101 **"What makes lives worth living":** Stoknes, *What We Think about When We Try Not to Think about Global Warming,* pp. 114, 115.

p. 102 **"shift the balance":** Ibid., p. 121.

p. 102 **"What is an effective climate message?"** Ibid., p. 101.

p. 102 **"the Great Paradox":** Hochschild, *Strangers in Their Own Land,* p. 9.

p. 102 **"stewardship":** On framing climate change in terms of "stewardship" and "moral purity," see "Dialogue and Exchange."

p. 103 *New York Times* **profile:** Tabuchi, "In America's Heartland, Discussing Climate Change without Saying 'Climate Change.'" The following quotes by Miriam Horn are from this profile as well.

p. 103 **"The understanding of our interdependence":** Interview with Andrew Amelinckx, "Heroes of the Heartland: A Chat with Author Miriam Horn."

p. 103 **"Purity politics":** The politics of purity in the environmental movement is well documented and critiqued, and is related to the way disgust works in environmental politics (see note above). See, for example, Zimrig, *Clean and White;* Shotwell, *Against Purity;* Gamber, *Positive Pollutions and Cultural Toxins;* and Ray, *The Ecological Other.* Seymour is particularly interested in how a politics of purity prevent us from intersectional environmental advocacy that addresses issues of class, sexuality, and gender, the modes of which are often deliberately irreverence or disgust, to challenge "purity" as an environmental mode.

p. 104 **"existential exceptionalism":** Heglar, "Climate Change Ain't the First Existential Threat."

p. 105 **Green New Deal:** For the text of Ocasio-Cortez's resolution, see https://ocasio-cortez.house.gov/sites/ocasio-cortez.house.gov/files /Resolution%20on%20a%20Green%20New%20Deal.pdf.

p. 105 **Ocasio-Cortez's impassioned speech on climate change:** Available in part at www.theguardian.com/global/video/2019/mar/27 /people-are-dying-ocasio-cortez-delivers-fiery-speech-on-climate-inaction-video.

p. 107 **"It is because of empathy":** Bloom, *Against Empathy,* p. 127.

p. 107 **empathy as being central to both Nazi ideology:** Ibid., p. 159.

p. 108 **"it is more often":** Shuman, *Other People's Stories*, p. 5.

p. 109 **Compassion vs. empathy:** Sources I draw on are Tania Singer's groundbreaking research on neuroscience and compassion vs. empathy (starting with Singer et al., "Empathy for Pain Involves the Affective but Not Sensory Components of Pain"), Amy Shuman's *Other People's Stories*, Paul Bloom's *Against Empathy* (see particularly pages 138–39), Hochschild's *Strangers in Their Own Land* (discussions on "the empathy wall" and "practical empathy"), Brewer's analysis of compassion in *The Craving Mind*, Tara Brach's talk "What Is It Like Being You?," and research on the neuroscience of these emotions, as glossed, for example, by Tara Well, "Compassion Is Better Than Empathy: Neuroscience Explains Why."

p. 109 **Empathy fatigue:** Singer and Klimecki, "Empathy and Compassion."

p. 109 **Mindfulness expert Judson Brewer:** Brewer, *The Craving Mind*, p. 182.

p. 111 **"gap instinct":** Rosling, *Factfulness*, p. 39.

p. 111 **"I don't personally know anyone involved in young, right-of-center politics":** Oliver Milman, "The Young Republicans Breaking with Their Party over Climate Change."

p. 112 **"Speaking differently":** Rorty, *Contingency, Irony, and Solidarity*, p. 7.

6. Move Beyond Hope, Ditch Guilt, and Laugh More

p. 115 **one's identity grants unearned privilege, especially if one is white:** Robin DiAngelo's *White Fragility* is the foundational text on this topic. Stress about racial privilege can trigger a range of defenses, and in the context of a predominately white environmental movement is often expressed as guilt and shame. Environmental studies as a field has also been mostly white, although the demographics are changing, in no small part due to the climate justice movement. A great introduction to this concept is the "environmental race and class privilege knapsack," Gregory Mengel's green revision of Peggy McIntosh's "white privilege knapsack," available at www.pachamama.org/news/race-and-class-privilege-in-the-environmental-movement. The first challenge is to get more comfortable with white fragility so as to transform it into solidarity and "ally-ship". See also Kuttner, "Stop Wallowing in Your White Guilt and Start Doing Something for Racial Justice"; Milstein and Griego, "Environmental Privilege Walk: Unpacking the Invisible Knapsack"; and Utt, "True Solidarity." On guilt as a political affect, see Ladino, *Memorials Matter.*

p. 116 **"no one likes to feel bad"**: Seymour, *Bad Environmentalism,* p. 17.

p. 116 **"ecological footprint"**: The exercise, available at www.footprintnet work.org/our-work/ecological-footprint, is often used in introductory courses or workshops to show participants how their personal habits affect the Earth.

p. 117 **"just fallen off the back of Santa's sleigh"**: The quotations in this paragraph are from Kimmerer, *Braiding Sweetgrass,* pp. 199, 179, and 183. See also Kimmerer's August 2012 TED Talk, "Reclaiming the Honorable Harvest," at www.youtube.com/watch?v=Lz1vgfZ3etE.

p. 118 **personal sacrifice won't matter**: Much has been written on this issue. See, for example, Maniates, "Teaching for Turbulence"; and Jensen, "Forget Shorter Showers." Additionally, the research areas of conservation psychology and environmental ethics are preoccupied with the question, On what basis do people make decisions to change behavior? However, as argued in chapter 3, concerns such as whether or not to buy a pen still center the individual (defined as a consumer) as the domain of action rather than the collective (as a polis, rather than just a collection of consumers). For more analysis along these lines, see Lukacs, "Neoliberalism Has Conned Us into Fighting Climate Change as Individuals."

p. 118 **"Make justice and liberation feel good"**: brown, *Pleasure Activism.*

p. 120 **social media addiction**: Twenge, *iGen;* Chowdry, "Research Links Heavy Facebook and Social Media Usage to Depression."

p. 121 **"Sustain the ones who sustain you"**: Kimmerer, *Braiding Sweetgrass,* p. 183.

p. 121 **"Hope is not the conviction"**: Havel, *Disturbing the Peace,* pp. 181–82; and www.vhlf.org/havel-quotes/disturbing-the-peace.

p. 121 **Havel's approach to hope differs**: My thinking about the concept of hope in this section is drawn from von Mossner, *Affective Ecologies;* Duncan-Andrade, "Note to Educators"; Jensen, "Beyond Hope"; McKibben, *Hope, Human and Wild;* Nairn, "Learning from Young People

Engaged in Climate Activism"; and Solnit, *Hope in the Dark* and "'Hope Is an Embrace of the Unknown.'" In *Living in Denial,* Kari Norgaard argues that hope is a kind of denial of problems, while environmental philosopher Tom Bristow adds that it can also become a distraction; see Bristow, Muir, and van Dooren, "Hope in a Time of Crisis." In the same article, Thom van Dooren calls hope a "passive abdication that says everything is okay, our consciences are assuaged, there's no need to think about responsibility, someone else has it covered." Hope "eschews reality" and "becomes self-censorship" that "serves the status quo." Heather Houser, in *Ecosickness in Contemporary U.S. Fiction,* counts the reasons we should be wary of the distinctly American fetish for "positive thinking": "First, optimism prevents us from recognizing signs of adversity for which we could prepare." Second, it supports "the capitalist mandate to grow at all costs" and "obfuscates 'the crueler aspects of the market economy' in favor of 'a harsh insistence on personal responsibility'" (pp. 220–21).

p. 121 **"principle of hope"**: von Mossner, *Affective Ecologies,* p. 163.

p. 122 **Swedish youth activist Greta Thunberg:** www.ted.com/talks /greta_thunberg_the_disarming_case_to_act_right_now_on_climate.

p. 122 **What we need is *critical* hope:** Duncan-Andrade, "Note to Educators," p. 191.

p. 123 **"future worlds"**: von Mossner, *Affective Ecologies,* p. 163.

p. 123 **"the deepest pleasure comes from"**: brown, *Pleasure Activism,* p. 14.

p. 124 **"reframing climate science"**: Haltinner and Sarathchandra, "Climate Change Skepticism as a Psychological Coping Strategy," p. 7.

p. 124 **"misery resistance"**: brown, *Emergent Strategy,* p. 33 (emphasis added).

p. 124 **humor. . . as a "resiliency skill"**: See, for example, Davenport, *Emotional Resiliency in the Era of Climate Change,* p. 135.

p. 125 **"It's as if giving up on saving the world"**: Eisenstein, *Climate—A New Story* (p. 74). See also Seymour, *Bad Environmentalism;* and Tara Brach's guided meditation "Letting Go into Living Presence" at www .tarabrach.com/meditation-letting-go-living-presence.

p. 125 **"It's at just this moment of crisis"**: Scranton, *We're Doomed. Now What?,* p. 7.

p. 125 **"a willingness to live without hope"**: Jamail, *The End of Ice,* p. 219.

p. 126 **savior complex:** See Cole, "The White-Savior Industrial Complex," about white people trying to "help" others in ways that aggrandize themselves at the cost of helping.

p. 126 **politics of sacrifice:** Whether guilt and sacrifice generate meaningful behavioral change is debated and varies across religions, cultures, and communities. Guilt registers differently across these identity categories, as we have seen. Some still argue that guilt is necessary as a stage in environmental consciousness-raising that will lead to sacrifice (I disagree), while others see sacrifice as a misguided way to individualize a problem that has broader structural causes. For a defense of sacrifice, see Meyer and Maniates (eds.), *The Environmental Politics of Sacrifice.* As mentioned regarding white fragility, environmentalism's emphasis on sacrifice may reflect its elitist proclivities: for one, it maintains this notion of the individual as the arena of action, and it assumes that a person has enough power or goods to sacrifice in the first place. As justice continues to influence the climate movement, the notions of pleasure, sacrifice, desire, and guilt will come to mean different things.

p. 126 **accept our own entanglement:** In *Against Purity,* Alexis Shotwell admonishes us not to adopt environmentalism's fetish for the natural and "pure," on the grounds that it erases the ways we are all complicit in structural injustice and gives us a sense of righteousness that ignores the myriad moral compromises we make daily.

7. Resist Burnout

p. 128 **'Transform yourself to transform the world':** Grace Lee Boggs, quoted in the dedication to brown, *Emergent Strategy.*

p. 128 **Self-care as activism:** Many of the sources I draw on for this book have step-by-step guidelines for how to do the work of self-study and -care. Because they focus specifically on the value of self-care in climate and/ or activist work, I especially recommend Elena Aguilar's *Onward: Cultivating Emotional Resilience in Educators;* adrienne maree brown's *Emergent Strategy: Shaping Change, Changing Worlds;* Leslie Davenport's *Emotional Resiliency in an Era of Climate Change;* Laura van Dernoot Lipsky's *The Age of Overwhelm: Strategies for the Long Haul;* and Alessandra Pigni's *The Idealist's Survival Kit: 75 Simple Ways to Avoid*

Burnout. Sources that focus on self-care beyond the activist context include Boggs, *The Next American Revolution;* brown, *Pleasure Activism;* Doppelt, *Transformational Resilience;* Eisenstein, *Climate—A New Story;* Lorde, *A Burst of Light;* Lipsky, *Trauma Stewardship;* Macy and Johnstone, *Active Hope;* and Spark, "Mourning the Ghost."

p. 130 **"Caring for myself is not self-indulgence":** Audre Lorde, *A Burst of Light,* p. 130.

p. 130 **"In order to":** Aguilar, *Onward,* pp. 6–7.

p. 132 **some of my favorites.** A good starting point is the books recommended above. For practicing mindfulness, see Suzuki, *Zen Mind, Beginner's Mind;* Brewer, *The Craving Mind;* Kabat-Zinn, *Mindfulness for All;* and Brach, *Radical Acceptance, True Refuge,* and her website, www.tarabrach.com. Other books in the self-care genre can explain the benefits of taking care of your body by protecting your sleep and pursuing healthy eating habits. Drawing on the Alcoholics Anonymous model, yoga instructor Aimee Lewis-Reau and self-described "force of good in the world" (and environmental studies grad) LaUra Schmidt have created a twelve-step healing program for people suffering from climate grief, called the Good Grief Network (www.goodgriefnetwork.org). Environmental scientist Amy Spark and community educator Jodi Lammiman run eco-grief retreats through their organization, Refugia (www.refugiaretreats.com), taking up the kind of healing work and "great turning" that Joanna Macy and others popularized. Perhaps the best example of an organization emphasizing the synergy between self-care and social change is Movement Generation (movementgeneration.org), based in Oakland, California.

p. 132 **"Opting to pause and reflect":** Friedman, *Thank You for Being Late,* p. 4.

p. 133 **the four key components:** Himelstein, quoted in Schwartz, "Why Mindfulness and Trauma-Informed Teaching Don't Always Go Together."

Conclusion

p. 137 **"We can 'attend' to the drop-down menu's 'Latest Headlines'":** Princen, "A Letter to a (Composite) Student in Environmental Studies," p. 15.

p. 139   **"People call my generation "Generation Z""**: Margolin, https://www
.kekemagazine.com/2019/10/06/young-climate-activists-speak-at-u-
n-climate-action-summit/.

p. 139   **"The term *resilience,* however, is itself problematic"**: Sources that
helped me understand the racial and cultural politics of the term *resilience* include Gerald Vizenor's *Survivance: Narratives of Native Presence,*
Kyle Powys Whyte's "Indigenous Science (Fiction) for the Anthropocene" and "Is It Colonial Déjà Vu? Indigenous Peoples and Climate
Justice," Jaskiran Dhillon's "Indigenous Resurgence, Decolonization,
and Movements for Environmental Justice," and Paul Outka's "Environmentalism after Despair." The term has different connotations and
roots in other disciplines; for psychology, see Davenport, *Emotional
Resiliency in the Era of Climate Change,* and for ecology, see Doppelt,
Transformational Resilience.

p. 140   **"whose interests are being served"**: Dhillon, "Indigenous Resurgence, Decolonization, and Movements for Environmental Justice," p. 2.

p. 140   **"It is an extension of ongoing extinctions"**: Many anti-colonial scholars criticize climate crisis rhetoric for this reason, but here I'm drawing
on Whyte, "Indigenous Science (Fiction) for the Anthropocene."

p. 141   **"Resilience is a post-despair environmentalism"**: Outka, "Environmentalism after Despair," p. 2.

p. 142   **"resilience is by definition *against resistance*"**: Neocleous, "Resisting Resilience," p. 7.

p. 143   **"Tell me, what is it you plan to do"**: Mary Oliver, "The Summer Day."

Bibliography

Abrams, David. *The Spell of the Sensuous: Perception and Language in a More-Than-Human World*. New York: Vintage, 1997.

Aguilar, Elena. *Onward: Cultivating Emotional Resilience in Educators*. San Francisco: Jossey-Bass, 2018.

Ahearn, Ashley. "A Seattle Filmmaker Confronts His Grief Over a World That's Changing." *OPB*, May 2, 2017. www.opb.org/news/article /a-seattle-filmmaker-confronts-his-grief-over-a-world-thats-changing .

Ahmed, Sara. *The Promise of Happiness*. Durham, NC: Duke University Press, 2010.

Alaimo, Stacy. *Bodily Natures: Science, Environment, and the Material Self*. Bloomington: Indiana University Press, 2010.

———. *Exposed: Environmental Politics and Pleasures in Posthuman Times*. Minneapolis: University of Minnesota Press, 2016.

———. "The Trouble with Texts; or, Green Cultural Studies in Texas." In *Teaching North American Environmental Literature*, edited by Laird Christensen, Mark C. Long, and Fred Waage. New York: Modern Languages Association of America, 2008.

Albrecht, Glenn. *Earth Emotions: New Words for a New World*. Ithaca, NY: Cornell University Press, 2019.

Albrecht, Glenn, G. M. Sartore, L. Connor, N. Higginbotham, S. Freeman, B. Kelly, H. Stain, A. Tonna, and G. Pollard. "Solastalgia: The Distress Caused by Environmental Change." *Australas Psychiatry* 12, no. 10 (2007): 95–98.

Aldrich, Daniel, and Michelle A. Meyer. "Social Capital and Community Resilience." *American Behavioral Scientist* 59, no. 2 (2014): 254–69.

Amelinckx, Andrew. "Heroes of the Heartland: A Chat with Author Miriam Horn." *Modern Farmer,* September 29, 2016. https://modernfarmer.com /2016/09/miriam-horn.

Atkinson, Jennifer. "Addressing Climate Change Makes You a Badass, Not a Snowflake." *High Country News,* May 29, 2018.

"Attention, Please." *TED Radio Hour,* May 25, 2018. www.npr.org/programs /ted-radio-hour/614007696/attention-please

Barenberg, Otto, and Sofia Corzo. "Generation Z Is Not Afraid." *Harvard Political Review,* April 22, 2019. https://harvardpolitics.com/united-states /hpop-gen-z.

Barker, Pamela, and Amy McConnell Franklin. "Social and Emotional Learning for a Challenging Century." In *Earth Ed: Rethinking Education on a Changing Planet,* edited by Erik Assadourian. Washington, DC: Island Press, 2017.

Beck, Ulrich. *Risk Society: Toward a New Modernity.* London: Sage, 1992.

Bendell, Jem. "Deep Adaptation: A Map for Navigating Climate Tragedy." *IFLAS Occasional Paper 2,* July 27, 2018. Available at www.lifeworth.com /deepadaptation.pdf.

Bennett, Jane. *Vibrant Matter: A Political Ecology of Things.* Chapel Hill, NC: Duke University Press, 2010.

Bercovitch, Sacvan. *American Jeremiad.* Madison: University of Wisconsin Press, 1978.

Berg, Maggie, and Barbara Seeler. *The Slow Professor: Challenging the Culture of Speed in the Academy.* Toronto: University of Toronto Press, 2016.

Bergman, Carla, and Nick Montgomery. *Joyful Militancy: Building Thriving Resistance in Toxic Times.* Chico, CA: AK Press, 2017.

Berlant, Lauren. "Cruel Optimism." In *The Affect Theory Reader,* edited by Melissa Gregg and Gregory J. Seigworth. Durham, NC: Duke University Press, 2010.

Bigelow, Bill. "Teaching Climate Change." In *A People's Curriculum for the Earth: Teaching Climate Change and the Environmental Crisis,* edited by Bill Bigelow and Tim Swinehart. Milwaukee: Rethinking Schools Ltd., 2015.

Black, Toban, Tony Weis, Stephen d'Arcy, and Joshua Kahn Russell. *A Line in the Tar Sands: Struggles for Environmental Justice.* Oakland, CA: PM Press, 2014.

Bladow, Kyle, and Jennifer Ladino, eds. *Affective Ecocriticism: Emotion, Embodiment, Environment.* Lincoln: University of Nebraska Press, 2018.

Bloom, Paul. *Against Empathy: The Case for Radical Compassion.* New York: Ecco, 2016.

Boggs, Grace Lee. *The Next American Revolution: Sustainable Activism for the Twenty-First Century.* Berkeley: University of California Press, 2012.

Bourdieu, Pierre. *Distinction: A Social Critique of the Judgement of Taste.* Cambridge, MA: Harvard University Press, 1987.

Bowler, Peter. *Norton History of the Environmental Sciences.* New York: W.W. Norton, 1993.

Boyd, Andrew, and Dave Oswald Mitchell. *Beautiful Trouble: A Toolbox for Revolution.* New York: OR Books, 2016.

Brach, Tara. *Radical Acceptance: Embracing Your Life with the Heart of a Buddha.* New York: Bantam Dell, 2003.

———. *True Refuge: Finding Peace and Freedom in Your Own Awakened Heart.* New York: Bantam, 2012.

———. "What Is It Like Being You?" N.d. www.tarabrach.com/what-is-it-like-being-you.

Brecher, Jeremy. *Against Doom: A Climate Insurgency Manual.* Oakland, CA: PM Press, 2017.

Brewer, Judson. *The Craving Mind: From Cigarettes to Smartphones to Love—Why We Get Hooked and How We Can Break Bad Habits.* New Haven, CT: Yale University Press, 2018.

Bristow, Thom, Cameron Muir, and Thom van Dooren. "Hope in a Time of Crisis: Environmental Humanities and Histories of Emotions." *Histories of Emotion from Medieval Europe to Contemporary Australia,* November 6, 2015. https://historiesofemotion.com/2015/11/06/hope-in-a-time-of-crisis-environmental-humanities-and-histories-of-emotions.

brown, adrienne maree. *Emergent Strategy: Shaping Change, Changing Worlds.* Chico, CA: AK Press, 2017.

———. *Pleasure Activism: The Politics of Feeling Good.* Chico, CA: AK Press, 2019.

Bullard, Robert A. *Dumping in Dixie: Race, Class, and Environmental Quality.* 3rd ed. New York: Routledge, 2000.

Burke, Marshall, Felipe González, Patrick Baylis, Sam Heft-Neal, Ceren Baysan, Sanjay Basu, and Solomon Hsiang. "Higher Temperatures

Increase Suicide Rates in the United States and Mexico." *Nature Climate Change* 8 (2018): 723–29.

Burns, David. *Feeling Good: The New Mood Therapy.* New York: Harper, 2008.

Campbell, Joseph. *The Hero with a Thousand Faces.* 3rd ed. Novato, CA: New World Library, 2008.

Cavanagh, Sarah R. *The Spark of Learning: Energizing the College Classroom with the Science of Emotion.* Morgantown: West Virginia University Press, 2016.

Chowdry, Amit. "Research Links Heavy Facebook and Social Media Usage to Depression." *Forbes,* April 30, 2016.

Clark, Timothy. *Ecocriticism on the Edge: The Anthropocene as a Threshold Concept.* London: Bloomsbury Academic, 2015.

Clayton, Susan. "Mental Health Risk and Resilience among Climate Scientists." *Nature Climate Change* 8 (2018).

Clayton, Susan, Christie Manning, Kirra Krygsman, and Meighen Speiser, eds. *Mental Health and Our Changing Climate: Impacts, Implications, and Guidance.* Washington, DC: American Psychological Association and ecoAmerica, 2017. Available at https://www.apa.org/news/press/releases/2017/03/mental-health-climate.pdf.

Cohen, Steve. "The Age Gap in Environmental Politics." *State of the Planet,* February 4, 2019. https://blogs.ei.columbia.edu/2019/02/04/age-gap-environmental-politics.

Cole, Teju. "The White Savior Industrial Complex." *Atlantic,* March 21, 2012.

Commission for Racial Justice. *Toxic Wastes and Race in the United States: A National Report on the Racial and Socio-Economic Characteristics of Communities with Hazardous Waste Sites.* New York: Commission for Racial Justice, United Church of Christ, 1987. Available at http://d3n8a8pro7vhmx.cloudfront.net/unitedchurchofchrist/legacy_url/13567/toxwrace87.pdf?1418439935.

Corburn, Jason. *Street Science: Community Knowledge and Environmental Health Justice.* Cambridge, MA: MIT Press, 2005.

Correal, Annie. "What Drove a Man to Set Himself on Fire in Brooklyn?" *New York Times,* May 28, 2018.

Cortez, Marisol. "The Praxis of Deceleration: Recovery as 'Inner Work, Public Act.'" *Academic Labor: Research and Artistry* 2 (2018).

Cronon, William. "A Place for Stories: Nature, History, Narrative." *Journal of American History* 78, no. 4 (1992): 1347–76.

Crutchfield, Rashida, and Jennifer Maguire. "Study of Student Basic Needs: California State University Basic Needs Initiative." January 2018. https://www2.calstate.edu/impact-of-the-csu/student-success/basic-needs-initiative/Documents/BasicNeedsStudy_phaseII_withAccessibility Comments.pdf

Crutzen, Paul J., and Eugene F. Stoermer. "The 'Anthropocene.'" *IGBP Newsletter,* no. 41 (May 2000): 17–18.

Cunsolo, Ashlee, and Neville Ellis. "Ecological Grief as a Mental Health Response to Climate Change–Related Loss." *Nature Climate Change* 8 (2018): 275–81.

Davenport, Leslie. *Emotional Resiliency in the Era of Climate Change: A Clinician's Guide.* London: Jessica Kingsley, 2017.

Davis, Diana K. *Resurrecting the Granary of Rome: Environmental History and French Colonial Expansion in North Africa.* Athens: Ohio University Press, 2007.

Demockar, Mary. *A Parent's Guide to Climate Revolution: 100 Ways to Build a Fossil-Free Future, Raise Empowered Kids, and Still Get a Good Night's Sleep.* Novato, CA: New World Library, 2018.

Dhillon, Jaskiran. "Indigenous Resurgence, Decolonization, and Movements for Environmental Justice." *Environment and Society* 9 (September 2018).

"Dialogue and Exchange." *TED Radio Hour,* October 27, 2017. www.npr.org/programs/ted-radio-hour/558307433.

DiAngelo, Robin. *White Fragility: Why It's So Hard for White People to Talk about Racism.* Boston: Beacon, 2018.

Di Chiro, Giovanna. "Environmental Justice." In *Keywords for Environmental Studies,* edited by Joni Adamson, William A. Geason, and David N. Pellow. New York: New York University Press, 2016. Available at https://keywords.nyupress.org/environmental-studies/essay/queer-ecology.

———. "Polluted Politics? Confronting Toxic Discourse, Sex Panic, and Eco-Normativity." In *Queer Ecologies: Sex, Nature, Politics, Desire,* edited by Cate Mortimer-Sandilands and Bruce Erickson. Bloomington: University of Indiana Press, 2010.

Doppelt, Bob. *Transformational Resilience: How Building Human Resilience to Climate Disruption Can Safeguard Society and Increase Wellbeing.* New York: Routledge, 2016.

Dowie, Mark. "Conservation Refugees." *Orion,* n.d. https://orionmagazine .org/article/conservation-refugees.

Ducharme, Jamie. "More Than 90% of Generation Z Is Stressed Out. And Gun Violence Is Partly to Blame." *Time,* October 30, 2018.

Duncan-Andrade, J. M. R. "Note to Educators: Hope Required When Growing Roses in Concrete." *Harvard Educational Review* 79, no. 2 (2009): 181–94.

Dunlap, Julie, and Susan A. Cohen, eds. *Coming of Age at the End of Nature: A Generation Faces Living on a Changed Planet.* San Antonio, TX: Trinity Press, 2016.

Dunne, Daisy. "Impact of Climate Change on Health Is 'the Major Threat of 21st Century.'" *CarbonBrief,* October 30, 2017. www.carbonbrief.org /impact-climate-change-health-is-major-threat-21st-century.

Easterbrook, Gregg. *It's Better Than It Looks: Reasons for Optimism in an Age of Fear.* New York: PublicAffairs, 2018.

Eaton, Marie. "Navigating Anger, Fear, Grief, and Despair." In *Contemplative Approaches to Sustainability in Higher Education: Theory and Practice,* edited by Marie Eaton, Holly J. Hughes, and Jean MacGregor. New York: Routledge, 2017.

Eaton, Marie, Holly J. Hughes, and Jean MacGregor, eds. *Contemplative Approaches to Sustainability in Higher Education: Theory and Practice.* New York: Routledge, 2017.

Eckersley, Richard. "Nihilism, Fundamentalism, or Activism: Three Responses to Fears of the Apocalypse." *The Futurist* 42, no. 1 (2008): 35–39.

Eisenstein, Charles. *Climate—A New Story.* Berkeley, CA: North Atlantic Books, 2018.

"Elin Kelsey on Contagious Hope, Children's Books, and Solutions." *Big Planet, Big Feels,* March 4, 2019. https://www.bigplanetbigfeels.net /episodes/episode-6-elin-kelsey-on-contagious-hope-childrens-books-and-solutions.

Fiskio, Janet. "Building Paradise in the Classroom." In *Teaching Climate Change in the Humanities,* edited by Stephen Siperstein, Shane Hall, and Stephanie Lemenager. New York: Routledge, 2016.

Florida, Richard. "The Unhappy States of America." *CityLab,* March 20, 2018. www.citylab.com/life/2018/03/the-unhappy-states-of-america/555800.

Follmann, Clare. "The Art of Arguing Science: A Critique of Scientific Language and Rhetoric through the Invasive Species Narrative." Master's thesis, Evergreen State College, Olympia, WA, June 2018.

Ford, Allison, and Kari Marie Norgaard. "From Denial to Resistance: How Emotions and Culture Shape Our Responses to Climate Change." In *Climate and Culture: Multidisciplinary Perspectives on a Warming World,* edited by Giuseppe Feola, Hilary Geoghegan, and Alex Arnall, 219–42. Cambridge: Cambridge University Press, 2019.

Franzen, Jonathan. *The End of the End of the Earth.* London: 4th Estate, 2018.

———. "What If We Stopped Pretending?" *New Yorker,* September 8, 2019.

Freire, Paulo. *Pedagogy of the Oppressed.* New York: Herder & Herder, 1972.

Friedman, Thomas L. *Thank You for Being Late: An Optimist's Guide to Thriving in an Age of Accelerations.* New York: Picador, 2016.

Fritze, Jessica G., Grant A. Blashki, Susie Burke, and John Wiseman. "Hope, Despair, and Transformation: Climate Change and the Promotion of Mental Health and Wellbeing." *International Journal of Mental Health Systems* 2 (September 2008): 1–10.

Funk, Cary, and Meg Hefferon. "Many Republican Millennials Differ with Older Members on Climate Change and Energy Issues." Pew Research Center, May 14, 2018. www.pewresearch.org/fact tank/2018/05/14 /many-republican-millennials-differ-with-older-party-members-on-climate-change-and-energy-issues.

Gabney, Bruce Cannon. *A Generation of Sociopaths: How the Baby Boomers Betrayed America.* New York: Hachette Books, 2017.

Gamber, John Blair. *Positive Pollutions and Cultural Toxins: Waste and Contamination in Contemporary U.S. Ethnic Literatures.* Lincoln: University of Nebraska Press, 2012.

Garrard, Greg. *Ecocriticism.* 2nd ed. New York: Routledge, 2012.

"Generation Z Stressed about Issues in the News but Least Likely to Vote." *ScienceDaily,* October 30, 2018. www.sciencedaily.com/releases/2018/10 /181030093709.htm

Germic, Stephen. *American Green: Class, Crisis, and the Deployment of Nature in Central Park, Yosemite, and Yellowstone.* Lanham, MD: Lexington Books, 2001.

Ghosh, Amitav. *The Great Derangement: Climate Change and the Unthinkable.* Chicago: University of Chicago Press, 2017.

"Global Warming's Six Americas." Yale Program on Climate Change Communication, November 1, 2016. https://climatecommunication.yale .edu/about/projects/global-warmings-six-americas

Goldrick-Rab, Sara, and Jesse Stommel. "Teaching the Students We Have, Not the Students We Wish We Had." *Chronicle of Higher Education,* December 21, 2018.

Goleman, Daniel. *Emotional Intelligence: Why It Can Matter More Than IQ.* New York: Bantam, 1995.

Green, Emily. "The Existential Dread of Climate Change." *Psychology Today,* October 13, 2017.

"A Green New Deal Must Be Rooted in a Just Transition for Workers and Communities Most Impacted by Climate Change." Climate Justice Alliance, December 10, 2018. https://climatejusticealliance.org/green-new-deal-must-rooted-just-transition-workers-communities-impacted-climate-change.

Grove, Richard. *Green Imperialism: Colonial Expansion, Tropical Island Edens, and the Origins of Environmentalism 1600–1860.* Cambridge: Cambridge University Press, 1996.

Haiven, Max, and Alex Khasnabish. *The Radical Imagination.* London: Zed Books, 2014.

Haltinner, Kristin, and Dilshani Sarathchandra. "Climate Change Skepticism as a Psychological Coping Strategy." *Sociology Compass* 12, no. 6 (2018): 1–10.

Hanh, Thich Nhat. *The Art of Living: Peace and Freedom in the Here and Now.* New York: HarperOne, 2017.

Haraway, Donna. *Staying with the Trouble: Making Kin in the Chthulucene.* Chapel Hill, NC: Duke University Press, 2016.

Harding, Sandra. *Sciences from Below: Feminisms, Postcolonialities, and Modernities.* Chapel Hill, NC: Duke University Press, 2008.

———. *Whose Science? Whose Knowledge? Thinking from Women's Lives.* Ithaca: Cornell University Press, 1991.

Havel, Václav. *Disturbing the Peace: A Conversation with Karel Hvížďala.* New York: Vintage Books, 1991.

Hawken, Paul. *Blessed Unrest: How the Largest Movement in the World Came into Being and Why No One Saw It Coming.* London: Viking Penguin, 2007.

Hayhoe, Katharine. "The Most Important Thing You Can Do to Fight Climate Change: Talk about It." *TEDWomen,* November 2018. www.ted.com /talks/katharine_hayhoe_the_most_important_thing_you_can_do_to_fight_ climate_change_talk_about_it.

Heglar, Mary Annaïse. "Climate Change Ain't the First Existential Threat." *Medium,* February 18, 2019. https://medium.com/s/story/sorry-yall-but-climate-change-ain-t-the-first-existential-threat-b3c999267aa0.

Heise, Ursula. *Sense of Place and Sense of Planet: The Environmental Imagination of the Global.* Oxford: Oxford University Press, 2008.

Hochschild, Arlie. *Strangers in Their Own Land: Anger and Mourning on the American Right.* New York: New Press, 2018.

Hoffman, Andrew J. *How Culture Shapes the Climate Change Debate.* Palo Alto, CA: Stanford University Press, 2015.

hooks, bell. *Teaching Community: A Pedagogy of Hope.* New York: Routledge, 2003.

Horn, Miriam. *Rancher, Farmer, Fisherman: Conservation Heroes of the American Heartland.* New York: W. W. Norton, 2016.

Houser, Heather. *Ecosickness in Contemporary U.S. Fiction: Environment and Affect.* New York: Columbia University Press, 2014.

———. *Environmental Culture of the Infowhelm.* Forthcoming.

Hufnagel, Elizabeth. "Students' Emotional Connections to Climate Change: A Framework for Teaching and Learning." In *Teaching and Learning about Climate Change: A Framework for Educators,* edited by Daniel P. Shepardson, A. Roychoudhury, and Andrew S. Hirsch. New York: Routledge, 2017.

Hunt, Elle. "BirthStrikers: Meet the Women Who Refuse to Have Children until Climate Change Ends." *Guardian,* March 12, 2019.

An Inconvenient Truth. Directed by Davis Guggenheim. DVD, Lawrence Bender Productions, 2006.

"Inheriting Climate Change: What Will Boomers Leave Behind?" *Climate One,* May 8, 2017. https://climateone.org/audio/inheriting-climate-change.

Institute for College Access and Success. "Student Debt and the Class of 2018." September 19, 2019. https://ticas.org/affordability-2/student-aid /student-debt-student-aid/student-debt-and-the-class-of-2018.

Irfan, Umair. "We Need to Talk about the Ethics of Having Children in a Warming World." *Vox,* March 11, 2019. www.vox.com/2019/3/11/18256166 /climate-change-having-kids.

Jamail, Dahr. *The End of Ice: Bearing Witness and Finding Meaning in the Path of Climate Disruption*. New York: New Press, 2019.

Jamail, Dahr, and Barbara Cecil. "Rethink Activism in the Face of Catastrophic Biological Collapse." *Truthout*, March 4, 2019. https://truthout .org/articles/climate-collapse-is-on-the-horizon-we-must-act-anyway.

Jennings, Pilar. "Boundaries Make Good Bodhisattvas." *Tricycle*, Summer 2018.

Jensen, Derek. "Beyond Hope." *Orion*, May/June 2006. https://orionmagazine .org/article/beyond-hope/.

———. "Forget Shorter Showers." *Orion*, July 2009. https://orionmagazine .org/article/forget-shorter-showers.

Jones, Van. "Breaking Out of Our Resistance Bubble." *Sounds True* event and podcast, *Waking Up in the World*, September 24–October 3, 2018. Available at www.mixcloud.com/soundstrueinsightsattheedge/van-jones-breaking-out-of-our-resistance-bubble.

June, Aubrey Williams. "The Invisible Labor of Minority Professors." *Chronicle of Higher Education*, November 8, 2015.

Kabat-Zinn, Jon. *Mindfulness for All: The Wisdom to Transform the World*. New York: Hachette, 2019.

———. *Wherever You Go, There You Are: Mindfulness Meditation in Everyday Life*. New York: Hyperion, 1994.

Kahneman, Daniel. *Thinking, Fast and Slow*. New York: Farrar, Straus & Giroux, 2011.

Kalmus, Peter. *Being the Change: Live Well and Spark a Climate Revolution*. Gabriola Island, BC: New Society Publishers, 2017.

Kaufman, Michael. *The Time Has Come: Why Men Must Join the Gender Equality Revolution*. Berkeley, CA: Counterpoint, 2019.

Kelsey, Elin. "Birth of Ocean Optimism." www.oceanoptimism.org/birth-ocean-optimism.

———. "Introduction: 'Dear _____'—Writing Our Way beyond Doom and Gloom." In *Beyond Doom and Gloom: An Exploration through Letters*, edited by Elin Kelsey, *RCC Perspectives: Transformations in Environment and Society*, 2014, no. 6. Available at doi.org/10.5282/rcc/6804.

———. *Radical Hope: Inspiring Sustainability Transformations through Our Past—A Group-Sourced Syllabus*. N.d. Available at https://radicalhopesyllabus.com.

Kevorkian, Kriss A. "Environmental Grief: Hope and Healing." N.d. Available at https://citeseerx.ist.psu.edu/viewdoc/download?doi=10.1.1.561.289&rep=rep1&type=pdf.

Kimmerer, Robin Wall. *Braiding Sweetgrass: Indigenous Wisdom, Scientific Knowledge, and the Teachings of Plants.* Minneapolis: Milkweed Editions, 2015.

Klein, Naomi. *This Changes Everything: Capitalism vs. the Climate.* New York: Simon & Schuster, 2015.

Kohn, Sally, and Erick Erickson. "Relationship across Rupture." *On Being,* October 11, 2018. https://onbeing.org/programs/sally-kohn-and-erick-erickson-relationship-across-rupture-oct18.

Kolbert, Elizabeth. "How to Write About a Vanishing World." *New Yorker,* October 15, 2018.

———. *The Sixth Extinction: An Unnatural History.* New York: Henry Holt, 2014.

Kuttner, Robert. "Stop Wallowing in Your White Guilt and Start Doing Something for Racial Justice." *American Prospect,* August 21, 2018.

Ladino, Jennifer. *Memorials Matter: Emotion, Environment, and Public Memory at American Historical Sites.* Reno: University of Nevada Press, 2019.

Latour, Bruno. *We Have Never Been Modern.* Cambridge, MA: Harvard University Press, 1993.

Leiserowitz, Anthony, Edward Maibach, Seth Rosenthal, John Kotcher, Parrish Bergquist, Matthew Ballew, Matthew Goldberg, and Abel Gustafson. *Climate Change in the American Mind: April 2019.* Yale University and George Mason University. New Haven, CT: Yale Program on Climate Change Communication, 2019. Available at https://climate-communication.yale.edu/publications/climate-change-in-the-american-mind-april-2019.

Lilla, Mark. *The Once and Future Liberal: After Identity Politics.* New York: Harper, 2017.

Lipsky, Laura van Dernoot. *The Age of Overwhelm: Strategies for the Long Haul.* Oakland, CA: Berrett-Koehler, 2018.

———. *Trauma Stewardship: An Everyday Guide to Caring for Self while Caring for Others.* San Francisco: Berrett-Koehler, 2009.

Loeb, Paul Rogat, ed. *The Impossible Will Take a Little While: Perseverance and Hope in Troubled Times.* New York: Basic Books, 2014.

———. *Soul of a Citizen: Living with Conviction in Challenging Times.* New York: St. Martin's Press, 2010.

The Lorax. Directed by Christ Renaud, Cinco Paul, Ken Daurio, John Powell, Chris Meledandri, Audrey S. Geisel, Janet Healy et al. DVD, Universal Studios Home Entertainment, 2012.

Lorde, Audre. *A Burst of Light and Other Essays.* Ithaca, NY: Firebrand Books, 1988.

Lukacs, Martin. "Neoliberalism Has Conned Us into Fighting Climate Change as Individuals." *Guardian,* July 17, 2017.

Lukianoff, Greg, and Jonathan Haidt. "The Coddling of the American Mind." *Atlantic,* September 2015. www.theatlantic.com/magazine/archive/2015 /09/the-coddling-of-the-american-mind/399356/

———. *The Coddling of the American Mind: How Good Intentions and Bad Ideas Are Setting Up a Generation for Failure.* New York: Penguin, 2018.

Lythcott-Haims, Julie. *How to Raise an Adult: Break Free of the Overparenting Trap and Prepare Your Kid for Success.* New York: St. Martin's Griffin, 2016.

Macy, Joanna, and Chris Johnstone. *Active Hope: How to Face the Mess We're in without Going Crazy.* Novato, CA: New World Library, 2012.

Majeed, Haris and Jonathan Lee. "The Impact of Climate Change on Youth Depression and Mental Health." *The Lancet Planetary Health* 1 (June 2017).

Maniates, Michael. "Teaching for Turbulence." In *State of the World 2013: Is Sustainability Still Possible?,* edited by Worldwatch Institute. Washington, DC: Island Press, 2013.

Margulis, Lynn. *Symbiotic Planet: A New Look at Evolution.* New York: Basic Books, 1999.

Marshall, George. *Don't Even Think about It: Why Our Brains Are Wired to Ignore Climate Change.* New York: Bloomsbury, 2015.

Masson-Delmotte, V., P. Zhai, H.O. Pörtner, D. Roberts, J. Skea, P.R. Shukla, A. Pirani, W. Moufouma-Okia, C. Péan, R. Pidcock, S. Connors, J.B.R. Matthews, Y. Chen, X. Zhou, M.I. Gomis, E. Lonnoy, T. Maycock, M. Tignor, and T. Waterfield, eds. "Global Warming of 1.5°C: An IPCC Special Report on the Impacts of Global Warming of 1.5°C above Pre-Industrial Levels and Related Global Greenhouse Gas Emission Pathways, in the Context of Strengthening the Global Response to the

Threat of Climate Change, Sustainable Development, and Efforts to
Eradicate Poverty." Intergovernmental Panel on Climate Change, 2018.
Available at https://report.ipcc.ch/sr15/pdf/sr15_spm_final.pdf.

Mauch, Christof. "Slow Hope: Rethinking Ecologies of Crisis and Fear."
RCC Perspectives: Transformations in Environment and Society, 2019, no. 1:
1–43.

McDermott, Brian, dir. *Marathon for Justice.* EmpathyWorks Films, 2016.

McKibben, Bill. *The End of Nature.* New York: Random House, 1989.

———. *Hope, Human and Wild: True Stories of Living Lightly on the Earth.*
Minneapolis: Milkweed Editions, 2007.

Meadows, Donella. "Leverage Points: Places to Intervene in a System." *The
Donella Meadows Project.* http://donellameadows.org/archives/leverage-
points-places-to-intervene-in-a-system.

Mengel, Gregory. "Race and Class Privilege in the Environmental Movement." *Pachamama,* May 10, 2016. www.pachamama.org/news/race-and-
class-privilege-in-the-environmental-movement.

"Mental Health Issues in Students." *Chronicle of Higher Education,* October
27, 2016.

Merchant, Carolyn. *The Death of Nature: Women, Ecology, and the Scientific
Revolution.* New York: HarperCollins, 1980.

Meyer, John. *Engaging the Everyday: Environmental Social Criticism and the
Resonance Dilemma.* Cambridge, MA: MIT Press, 2015.

Meyer, John, and Michael Maniates, eds. *The Environmental Politics of
Sacrifice.* Cambridge, MA: MIT Press, 2010.

Milman, Oliver. "The Young Republicans Breaking with Their Party over
Climate Change." *Guardian,* April 12, 2019.

Milstein, Tima, and Stephen Griego. "Environmental Privilege Walk:
Unpacking the Invisible Knapsack." In *Environmental Communication
Pedagogy and Practice,* edited by Tema Milstein, Mairi Pileggi, and Eric L.
Morgan. London: Routledge, 2017.

Mineo, Liz. "Harvard Study, Almost 80 Years Old, Has Proved That Embracing Community Helps Us Live Longer, and Be Happier." *Harvard Gazette,*
April 11, 2017.

Morton, Timothy. "Don't Just Do Something, Sit There! Global Warming and
Ideology." In *Rethink: Contemporary Art and Climate Change,* edited by
Anne Sophie Witzke. Copenhagen: Alexandra Institute, 2009.

———. *The Ecological Thought.* Cambridge, MA: Harvard University Press, 2012.

———. *Hyperobjects: Philosophy and Ecology after the End of the World.* Minneapolis: University of Minnesota Press, 2013.

Mountz, Alison, Anne Bonds, Becky Mansfield, Jenna Loyd, Jennifer Hyndman, Margaret Walton-Roberts, Ranu Basu, Risa Whitson, Roberta Hawkins, Trina Hamilton, and Winifred Curran. "For Slow Scholarship: A Feminist Politics of Resistance through Collective Action in the Neoliberal University." *ACME: An International E-Journal for Critical Geographies* 14, no. 4 (2015): 1235–59.

Nairn, Karen. "Learning from Young People Engaged in Climate Activism: The Potential of Collectivizing Despair and Hope." *Young* (2019). 1–16.

Neocluous, Mark. "Resisting Resilience." *Radical Philosophy,* no. 178 (March /April 2013): 2–7.

"The Next Generations: We Can't Save the World without Them." *Global GoalsCast,* September 27, 2018. https://globalgoalscast.org/youth.

Nixon, Rob. *Slow Violence and the Environmentalism of the Poor.* Cambridge, MA: Harvard University Press, 2013.

Norgaard, Kari. "Climate Change Is a Social Issue." *Chronicle of Higher Education,* January 17, 2016. www.chronicle.com/article/Climate-Change-Is-a-Social/234908.

———. *Living in Denial: Climate Change, Emotions, and Everyday Life.* Cambridge, MA: MIT Press, 2011.

———. "The Sociological Imagination in a Time of Climate Change." *Global and Planetary Change* 163 (April 2018): 171–76.

Nussbaum, Martha. *Political Emotions: Why Love Matters for Justice.* Cambridge, MA: Belknap Press, 2013.

Oliver, Mary. "The Summer Day." In *New and Selected Poems,* p. 94. Boston: Beacon Press, 1992. Also available at www.loc.gov/poetry/180/133.html.

Orr, Deborah. "The Uses of Mindfulness in Anti-Oppressive Pedagogies: Philosophy and Praxis." *Canadian Journal of Education* 27, no. 4 (2002).

Outka, Paul. "Environmentalism after Despair." *Resilience: A Journal of the Environmental Humanities* 1, no. 1 (September 15, 2014).

Paris, Django, and H. Samy Alim, eds. *Culturally Sustaining Pedagogies: Teaching and Learning for Justice in a Changing World.* New York: Teachers College Press, 2017.

Pearl, Mike. "Climate Change Edgelords Are the New Climate Change Deniers." *Vice,* October 10, 2018.

Pellow, David Naguib. *What Is Critical Environmental Justice?* Cambridge: Polity, 2017.

Petersen, Anne Helen. "How Millennials Became the Burnout Generation." *BuzzFeed News,* January 5, 2019. www.buzzfeednews .com/article/annehelenpetersen/millennials-burnout-generation-debt-work.

Pigni, Alessandra. *The Idealist's Survival Kit: 75 Simple Ways to Avoid Burnout.* Berkeley, CA: Parallax Press, 2016.

Pilkington, Ed, and Mona Chalabi. "Climate Change: The Missing Issue of the 2016 Campaign." *Guardian,* July 5, 2016.

Pinker, Steven. *Enlightenment Now: The Case for Reason, Science, Humanism, and Progress.* New York: Viking, 2018.

———. "The Media Exaggerates Negative News: This Distortion Has Consequences." *Guardian,* February 17, 2018.

Poerksen, Uwe. *Plastic Words: The Tyranny of a Modular Language.* Translated by Jutta Mason and David Cayley. University Park: Pennsylvania State University Press, 1995.

Popescu, Irina. "The Educational Power of Discomfort." *Chronicle of Higher Education Online,* April 17, 2016. www.chronicle.com/article/The-Educational-Power-of/236136.

Princen, Thomas. "A Letter to a (Composite) Student in Environmental Studies." In *Beyond Doom and Gloom: An Exploration through Letters,* edited by Elin Kelsey. *RCC Perspectives,* 2014, no. 6.

Putnam, Robert. *Bowling Alone: The Collapse and Revival of American Community.* New York: Simon & Schuster, 2000.

Quiroz-Martinez, Julie, Diana Pei Wu, and Kristen Zimmerman. *ReGeneration: Young People Shaping Environmental Justice.* Oakland, CA: Movement Strategy Center, 2005.

Ragoonaden, Karen, ed. *Mindful Teaching and Learning: Developing a Pedagogy of Well-Being.* Lanham, MD: Lexington Books, 2017.

Rainey, James. "For Some Millennials, Climate Change Clock Ticks Louder Than Biological One." *MSNBC News,* April 21, 2019. www.nbcnews.com /news/us-news/some-millennials-climate-change-clock-ticks-louder-biological-one-n993331.

Ray, Sarah Jaquette. "Coming of Age at the End of the World: The Affective Arc of Environmental Studies Curricula." In *Affective Ecocriticism: Emotion, Embodiment, and Environment,* edited by Kyle Bladow and Jennifer Ladino. Lincoln: University of Nebraska Press, 2018.

———. *The Ecological Other: Environmental Exclusion in American Culture.* Tucson: University of Arizona Press, 2013.

Reidy, Chris. "Interior Transformation on the Pathway to a Viable Future." *Journal of Futures Studies* 20, no. 3 (March 2016): 35–54.

Reilly, Kate. "Record Numbers of College Students Are Seeking Treatment for Depression and Anxiety." *Time,* March 19, 2018.

Reinhardt, R. J. "Global Warming Age Gap: Younger Americans Most Worried." *Gallup News,* May 11, 2018. https://news.gallup.com/poll/234314/global-warming-age-gap-younger-americans-worried.aspx.

Reyes, Abigail. "Student Leadership Institute for Climate Resilience Curriculum Manual 3.0." University of California–Irvine, 2019.

Roos, David. "Are Millennials Really the First Generation to Do Worse Than Their Parents?" N.d. https://money.howstuffworks.com/personal-finance/financial-planning/millennials-first-worse-parents.htm.

Ropeik, David. "It's Time for Climate Change Communicators to Listen to Social Science." *Grist,* March 17, 2019. https://grist.org/article/its-time-for-climate-change-communicators-to-listen-to-social-science.

Rorty, Richard. *Contingency, Irony, and Solidarity.* Cambridge: Cambridge University Press, 1989.

Rosenberg, Brian. "Actually, Academe Never Was All That Great." *Chronicle of Higher Education,* December 4, 2018.

———. "How We Can Help Students Survive an Age of Anxiety." *Chronicle of Higher Education,* April 26, 2018.

Rosling, Hans, with Anna Rosling Rönnlund and Ola Rosling. *Factfulness: Ten Reasons We're Wrong about the World—and Why Things Are Better Than You Think.* New York: Flatiron Books, 2018.

Sandilands, Catriona. "Queer Ecology." In *Keywords for Environmental Studies,* edited by Joni Adamson, William A. Geason, and David N. Pellow. New York: New York University Press, 2016. Available at https://keywords.nyupress.org/environmental-studies/essay/queer-ecology.

Scher, Avichai. "'Climate Grief': The Growing Emotional Toll of Climate Change." *NBC News,* December 24, 2018. www.nbcnews.com/health

/mental-health/climate-grief-growing-emotional-toll-climate-change-
n946751.

Schwartz, Katrina. "Why Mindfulness and Trauma-Informed Teaching
Don't Always Go Together." *Mind/Shift,* January 27, 2019. www.kqed.org
/mindshift/52881/why-mindfulness-and-trauma-informed-teaching-
dont-always-go-together.

Scranton, Roy. *Learning to Die in the Anthropocene: Reflections on the End of a
Civilization.* San Francisco: City Lights, 2015.

———. *We're Doomed. Now What? Essays on War and Climate Change.* New
York: Soho Press, 2018.

Senior, Jennifer. *All Joy and No Fun: The Paradox of Modern Parenthood.* New
York: Ecco, 2015.

Seymour, Nicole. *Bad Environmentalism: Irony and Irreverence in the Ecological
Age.* Minneapolis: University of Minnesota Press, 2018.

———. *Strange Natures: Futurity, Empathy, and the Queer Ecological Imagination.* Urbana: University of Illinois Press, 2013.

Shell, Ellen Ruppel. "College May Not Be Worth It Anymore." *New York
Times,* May 16, 2018.

Shotwell, Alexis. *Against Purity: Living Ethically in Compromised Times.*
Minneapolis: University of Minnesota Press, 2016.

Shuman, Amy. *Other People's Stories: Entitlement Claims and the Critique of
Empathy.* Champaign: University of Illinois Press, 2010.

Singer, Peter. "The Drowning Child and the Expanding Circle." *New
Internationalist,* April 1997. Available at www.utilitarian.net/singer/by
/199704--.htm.

Singer, Tania, and Olga M. Klimecki. "Empathy and Compassion." *Current
Biology* 24, no. 18 (September 22, 2014): R875–R878. Available at www
.sciencedirect.com/science/article/pii/S0960982214007702.

Singer, Tania, Ben Seymour, John O'Doherty, Holder Kaube, Raymond J.
Dolan, and Chris D. Frith. "Empathy for Pain Involves the Affective but
Not Sensory Components of Pain." *Science,* no. 2303 (February 20, 2004).

Siperstein, Stephen, Shane Hall, and Stephanie Lemenager, eds. *Teaching
Climate Change in the Humanities.* New York: Routledge, 2016.

Sliwa, Jim. "Climate Change's Toll on Mental Health." American Psychological Association, March 29, 2017. www.apa.org/news/press/releases/2017
/03/climate-mental-health.

Slotkin, Richard. *Gunfighter Nation: The Myth of the Frontier in Twentieth-Century America*. Norman: University of Oklahoma Press, 1998.

Slovic, Scott. *The Feeling of Risk: New Perspectives on Risk Perception*. New York: Earthscan, 2010.

———. *Going Away to Think: Engagement, Retreat, and Ecocritical Responsibility*. Reno: University of Nevada Press, 2008.

Slovic, Scott, and Paul Slovic. "The Arithmetic of Compassion." *New York Times,* December 4, 2015. www.nytimes.com/2015/12/06/opinion/the-arithmetic-of-compassion.html.

———, eds. *Numbers and Nerves: Information, Emotion, and Meaning in a World of Data*. Corvallis: Oregon State University Press, 2015.

Smith, Deirdre. "Why the Climate Movement Must Stand with Ferguson." *350,* August 20, 2014. https://350.org/how-racial-justice-is-integral-to-confronting-climate-crisis.

Smith, Linda Tuhiwai. *Decolonizing Methodologies: Research and Indigenous Peoples*. London: Zed Books, 2012.

Solnit, Rebecca. "Acts of Hope: Challenging Empire on the World Stage." *TomDispatch,* June 14, 2005. www.tomdispatch.com/post/3273.

———. *Hope in the Dark: Untold Histories, Wild Possibilities*. Chicago: Haymarket Books, 2016.

———. "'Hope Is an Embrace of the Unknown': Rebecca Solnit on Dark Times." *Guardian,* July 15, 2016.

———. *The Mother of All Questions*. Chicago: Haymarket Books, 2017.

———. *A Paradise Built in Hell: Extraordinary Communities That Arise in Disaster*. London: Penguin, 2010.

Spark, Amy. "Mourning the Ghost: Ecological Grief in the Ghost River Valley." Master's thesis, University of Edinburgh, 2016.

Spence, Mark David. *Dispossessing the Wilderness: Indian Removal and the Making of the National Parks*. Oxford: Oxford University Press, 2000.

Stafford, Tom. "Why Bad News Dominates the Headlines." *BBC News,* July 29, 2014. www.bbc.com/future/story/20140728-why-is-all-the-news-bad.

Stoknes, Per Espen. *What We Think about When We Try Not to Think about Global Warming: Toward a New Psychology of Climate Action*. Hartford, VT: Chelsea Green, 2015.

Suzuki, Shunryu. *Zen Mind, Beginner's Mind: Informal Talks on Zen Meditation and Practice.* Boulder, CO: Shambhala, 1973.

Swaminathan, Ashwin, Robyn M. Lucas, David Harley, and Anthony J. McMichael. "Will Global Climate Change Alter Fundamental Human Immune Reactivity: Implications for Child Health?" *Children* (Basel) 1, no. 3 (2014): 403–23.

Tabuchi, Hiroko. "In America's Heartland, Discussing Climate Change without Saying 'Climate Change.'" *New York Times,* January 28, 2017. www .nytimes.com/2017/01/28/business/energy-environment/navigating-climate-change-in-americas-heartland.html.

Teirstein, Zora. "Seing Red on Climate." *Grist,* January 28, 2108. https://grist .org/article/climate-change-isnt-just-for-democrats-anymore-meet-the-eco-right-republicans.

Thompson, Becky. *Teaching with Tenderness: Toward an Embodied Practice.* Champaign: University of Illinois Press, 2017.

Thomson, Jennifer. "A History of Climate Justice." *Solutions* 5, no. 2 (March 2014): 89–92.

Toadvine, Ted. "Six Myths of Interdisciplinarity." *Thinking Nature* 1 (2011).

Treleaven, David. *Trauma-Sensitive Mindfulness: Practices for Safe and Transformative Healing.* New York: W.W. Norton, 2018.

Twenge, Jean M. *iGen: Why Today's Super-Connected Kids Are Growing Up Less Rebellious, More Tolerant, Less Happy—and Completely Unprepared for Adulthood.* New York: Atria, 2017.

Utt, Jamie. "True Solidarity: Moving Past Privilege Guilt." *Everyday Feminism Online,* March 26, 2014. http://everydayfeminism.com/2014/03/moving-past-privilege-guilt.

Västfjäll, Daniel, Paul Slovic, and Marcus Mayorga. "Pseudoinefficacy and the Arithmetic of Compassion." In *Numbers and Nerves: Information, Emotion, and Meaning in a World of Data,* edited by Scott Slovic and Paul Slovic. Corvallis: Oregon State University Press, 2015.

Vizenor, Gerald. *Survivance: Narratives of Native Presence.* Lincoln: University of Nebraska Press, 2008.

von Mossner, Alexa Meik. *Affective Ecologies: Empathy, Emotion, and Environmental Narrative.* Columbus: Ohio State University Press, 2017.

Wallace-Wells, David. *The Uninhabitable Earth: Life after Warming.* New York: Tim Duggan Books, 2019.

Wapner, Paul. "Contemplative Environmental Studies: Pedagogy for Self and Planet." *Journal of Contemplative Inquiry* 3, no. 1 (2016).

Weaver, Sandra Long. "High Anxiety: Colleges Are Seeing an Increase in the Number of Students with Diagnosable Mental Illness, Anxiety, and Depression." *Chronicle of Higher Education,* March 18, 2016, 25–26.

Well, Tara. "Compassion Is Better Than Empathy. Neuroscience Explains Why." *The Clarity,* March 4, 2017. www.psychologytoday.com/us/blog /the-clarity/201703/compassion-is-better-empathy.

Weston, Anthony. *Mobilizing the Green Imagination: An Exuberant Manifesto.* Gabriola, BC: New Society, 2012.

Whaley, Madi. "The Beautiful Environmentalist: On the Real Food Movement and the Disciplined Body." *Writing at the End of the World,* January 5, 2018. http://writingattheendoftheworld.blogspot.com/2018/01/the-beautiful-environmentalist-on-real.html.

Whetham, Carl. "James Johnson, 'The Arithmetic of Compassion': Rethinking the Politics of Photography." December 9, 2016. https://carlwhetham .photo.blog/2016/12/09/james-johnson-the-arithmetic-of-compassion-rethinking-the-politics-of-photography.

Whyte, Kyle Powys. "Indigenous Science (Fiction) for the Anthropocene: Ancestral Dystopias and Fantasies of Climate Change Crises." *Environment and Planning E: Nature and Space* 1 (2018): 224–42.

———. "Is It Colonial Déjà Vu? Indigenous Peoples and Climate Justice." In *Humanities for the Environment: Integrating Knowledges, Forging New Constellations of Practice,* edited by Joni Adamson, M. Davis, and H. Huang. London: Earthscan, 2017.

Williams, Joseph P. "Fighting Food Insecurity on College Campuses." *U.S. News and World Report,* February 4, 2019. www.usnews.com/news /healthiest-communities/articles/2019-02-04/a-fight-against-food-insecurity-hunger-on-college-campuses.

Woodward, Aylin. "Millennials and Gen Z Are Finally Gaining Ground in the Climate Battle—Here Are the Signs They're Winning." *Business Insider,* April 2019.

———. "Who Has the Right to Declare the Urgency of Addressing Climate Change?" Talk given at University of California Santa Cruz, Feminist

Science Studies conference, "Indigeneity and Climate Justice," May 31, 2019.

Yoder, Kate. "An Evangelical Leader Calls Young Christians to Save the Planet." *Grist*, August 12, 2019. https://grist.org/article/this-evangelical-leader-calls-young-christians-to-save-the-planet.

Zimrig, Carl. A. *Clean and White: A History of Environmental Racism in the United States*. New York: New York University Press, 2015.

Zinn, Howard. "The Optimism of Uncertainty." In *The Impossible Will Take a Little While: Perseverance and Hope in Troubled Times,* edited by Paul Rogat Loeb. New York: Basic Books, 2014.

Index

Page references in *italics* refer to illustrations.

blame instinct, 85

Blessed Unrest (Hawken), 69–70

Bloom, Paul, 107

Boggs, Grace Lee, 128

Bozmoski, Alex, 36

Brach, Tara, 43, 47, 159n

Braiding Sweetgrass (Kimmerer), 116–17

brain research, 46–47

"Breaking Out of Our Resistance Bubbles" (Jones), 131

Brewer, Judson, 109

brown, adrienne maree, 1, 13, 31; on burnout, 128; on emergent strategy, 55, 69; *Emergent Strategy*, 10–11, 124, 128, 147n; on intentionality, 77; on multiple paths of resistance, 78; on pleasure, 118, 124; on visible action, 63

Buckel, David, 18, 19

Buddhism: on acceptance of death, 125; on beginner's mind, 49–50; on compassion, 109; on desire, 123; on mindfulness, 42, 44, 47; on right thinking, 53–54, 63

Bullard, Robert, 27

burnout, 40–41, 109, 128–36

Business as Usual narrative, 92

capitalism, 80, 81; and ecological issues, 11, 26

capitalist productivism, 41, 155n, 159n

carbon sequestration, 103

carbon tax, 36, 65

Cecil, Barbara, 53, 58

children and family planning, 22

Church of Euthanasia, 19

Clark, Timothy, 20

class elitism and environmentalism, 97–98, 167n

climate anxiety (term), 6

climate change (term), 103, 168n

climate change communication, 27, 34–38

climate change education, 11–14

Climate Change in the American Mind (Yale), 94

Climate Connection (NPR program), 19

climate generation, defined, 2–6

climate justice: emotions and work on, 18–29, 39–42; *vs.* environmentalism, 152n; links with social justice, 3, 5, 11, 23–28, 104–6

Climate Justice Alliance (CJA), 105

climate overwhelm, 10, 15, 34–35, 151n, 158n

Climate Psychological Alliance, 21–22

climate suicide, 18

climate wisdom, 30, 159n

coercive conservation, 62, 68, 161n

cognitive dissonance, 158n

cognitive psychology, 82–86

collective amnesia, 55–56

collective resilience, 7, 10, 28, 91, 135, 139. *See also* resilience

collectivity, 69–73, 133–35

colonialism, 5, 23, 62, 116

communication, 27, 34–38, 98–106. *See also* infowhelm; storytelling

compassion fatigue, 109, 154n

compassion *vs.* empathy, 95, 98, 106–13

confirmation bias, 26

emotions: climate justice work and, 18–29, 31–38, 157n; in learning environment, 157n; resilience of, 95–96; storytelling and, 92–95

empathy fatigue, 109

empathy *vs.* compassion, 95, 98, 106–13

empowerment, 8–11

The End of Ice (Jamail), 125

Enlightenment Now (Pinker), 88

Environmental Defense Fund, 6

environmental humanities, 8–9, 27, 150n

environmentalist disgust, 73, 97, 106–7, 167n

environmental justice movement. *See* climate justice

environmental studies curricula, 11–13

environmental trauma, 10, 28. *See also* emotional and physical trauma

epistemologies, 25–26

existential dread, 21–22, 141–42

extinction of nonhuman animal species, 32, 35, 140

Extinction Rebellion, 6

Factfulness (Rosling), 84–86, 88, 165n

facts and emotional responses, 34–39

family planning, 22

fear, 34, 82–86, 110–11. *See also* emotions

Finley, Ron, 52, 76

fires. *See* wildfires

5 Ds of climate communication, 36–38

fragility, 2, 147n, 170n, 173n. *See also* guilt

Franklin, Amy McConnell, 33–34

Franzen, Jonathan, 121

Friedman, Thomas, 132–33

frontier thesis, 68, 162n

Gandhi, 60

gap instinct, 84, 111

gardening, 52, 55, 76

generalization instinct, 85

generational differences on climate change, 7, 139, 148n. *See also* climate generation, defined

genocide, 23, 162n

"Global Warming's Six Americas" (Yale), 38

global youth climate strikes (2019), 8, 56, 106

Going Away to Think (Slovic), 67

Goldman, Emma, 119

Goleman, Daniel, 33

Gore, Al, 19, 36

The Great Turning narrative, 93, 94

The Great Unraveling narrative, 92–93

Green, Emily, 21–22

Green New Deal, 6, 23–24, 105–6

Green Tea Party, 102

grief, 120, 125–26. *See also* eco-grief; emotions

Guerra, Pia, *138*

guilt, 18, 40, 114–19, 173n. *See also* fragility

happiness, 28, 39–42

Havel, Václav, 121

Hawken, Paul, 69–70

Hayhoe, Katharine, 100

health of humans, 101–2

Heglar, Mary Annaïse, 104, 141–42

Herbert, Zbigniew, 74
Himelstein, Sam, 133
Hochschild, Arlie, 102
hope, 121–27, 172n. *See also* optimism
Hope, Human, and Wild (McKibben),
121
Hope in the Dark (Solnit), 121
Horn, Miriam, 103
Houser, Heather, 35
humanism, 24
Humboldt Bay eelgrass protection,
67
humor, 124
hurricanes, 4, 20, 82, 83, 149n

identity: climate change effects on,
21; communication and, 36, 37–38
imaginative solutions, 8–11
impermanence, 125–26
imposter syndrome, 54–55
indigenous peoples: effects of
colonialism and genocide on, 5,
23, 162n; identity and loss of
natural resources, 21; national
parks and effects on, 68–69, 162n;
on resilience, 139–40; sovereignty
of, 23–24; traditional ecological
knowledge of, 25–26
individual actions, 54–59, 63–66,
117–19, 171n
information deficit, as myth, 35–36
infowhelm, 34–35, 158n
infrastructure resilience, 57
insecurity, 34
instincts, 84–85
instrumentalism, 61–64, 122, 160n
interbeing, 73
interdependence, 69–73

IPCC (United Nations Intergovernmental Panel on Climate Change),
4, 149n
irony, 61, 124–25
It's Better Than It Looks (Easterbrook), 88

Jamail, Dahr, 53, 58, 125
Jensen, Derek, 122
jeremiad, 87, 165n
Johnstone, Chris, 92
Jones, Van, 131
Jordan, Chris, 18–19
joy, 119–21, 124
Just Transition collective, 139, 166n

Kabat-Zinn, Jon, 42–43
Kahneman, Daniel, 83
Kelsey, Elin, 19, 81, 162n
Kimmerer, Robin Wall, 72, 116–17, 121
King, Martin Luther, Jr., 55

Ladino, Jennifer, 158n
language. *See* communication;
storytelling
Laysan albatross, 18–19
leadership, 65–66
Learning to Die in the Anthropocene
(Scranton), 125
Le Guin, Ursula K., 81
Leiserowitz, Anthony, 82–83
Leunig, Michael, *8*
leverage points, 56–57
Lipsky, Laura van Dernoot, 10
listening actively, 98–99
Living in Denial (Norgaard), 35–36,
158n
localization, 100–101

Pacific Northwest indigenous tribes, 21
Palin, Sarah, 110
patriarchal notions of action, 63
Pei Wu, Diana, 65
Pellow, David N., 101
People's Climate March, 27
personal resilience, 13, 28, 58, 61, 71, 137–42, 161n. *See also* resilience; self-care
Pinker, Steven, 82, 83–84, 88
"The Place for Stories" (Cronon), 86–87
plastic pollution, 18–19, 101
pleasure, 118–19, 123–24
Pleasure Activism (brown), 124
politics: polarization in, 98–100, 111; of purity, 103–4, 169n; scientists involvement in, 32–33
pollution, 18–19, 101, 163n
post-traumatic stress disorder (PTSD), 20
poverty, 14, 28, 100, 105, 142, 167n. *See also* class elitism and environmentalism; social justice
powerlessness, 64, 73–76. *See also* resilience; small actions
practical reverence, 117
pre-traumatic stress (term), 6, 19
Princen, Thomas, 137
privilege, 114–15. *See also* elitism and environmentalism; guilt; social justice
productivism, capitalist, 41, 155n, 159n
progressive *vs.* declension narratives, 86–89
pseudoinefficacy, 73–75

psychoterratica, 20–21
pure, fetish of, 90, 126, 173n
purity politics, 103–4, 169n

queer ecological studies, 22–23
Quiroz-Martinez, Julie, 65

racial privilege, 114–15, 170n, 173n. *See also* elitism and environmentalism; guilt
racism, 11, 14, 68, 102, 141–42. *See also* social justice
RAIN process, 47, 159n
Rancher, Farmer, Fisherman (Horn), 103
reducing carbon emissions, 36, 57
refugee crisis, 62
ReGeneration (Quiroz-Martinez), 65, 161n
reproductive decisions, 22
Republican Party, 7
RepublicEn, 36
resilience, 28, 45, 137–43, 175n. *See also* collective resilience; personal resilience
resistance: multiple paths to, 78; resilience and, 141–43; self-care as, 128–35
resonance dilemma, 159n
"Rethink Activism in the Face of Catastrophic Biological Collapse" (Jamail and Cecil), 58
Reyes, Abigail, 1
right thinking, 53–54, 63
rising sea levels, 4
risk perception, 82–83, 91, 164n
Ron Finley Project, 52
Rorty, Richard, 112

Rosling, Hans, 84–86, 88, 111, 165n
rugged individualism, as myth, 71–72. *See also* individual actions

sacrifice, 40–42, 129–30
salmon, 21
savior complex, 126, 173n. *See also* guilt
science: critiques and use of, 25–26, 154n, 155n; emotional lives and, 33–39; political involvement and, 32–33
Scranton, Roy, 125
"Seashell" (Leunig), *8*
sea levels, 4
sea turtles, 116, 118
secondary grief (term), 6
self-care, 128–33, 174n; advocacy and, 133–36
self-erasure. *See* suicide and suicidal ideation
self-regulation, 46–47. *See also* mindfulness
sense of purpose, 28, 53, 143
Seymour, Nicole, 61, 63–64, 103, 116, 120, 123, 124
Shuman, Amy, 108
Singer, Peter, 74
Singer, Tania, 109
single-perspective instinct, 85
size instinct, 85
slavery, 141–42
Slovic, Paul, 36, 74, 75
Slovic, Scott, 36, 67
slow hope, 67–69
slow thinking, 66–68, 155n
slow violence, 5, 149n
Slow Violence (Nixon), 83, 149n

small actions, 54–59, 63–66. *See also* individual actions
snowflakes, as metaphor for youth, 2, 147n
social change agents, strategies for, 54–61
Social Darwinism, 25
social justice, 3, 5, 11, 23–28, 104–6, 152n
social movement theory, 28, 54, 59, 156n
social reproduction, 135
social sciences on emotional responses, 32–33
solastalgia, 6, 19, 21, 28, 152n
Solnit, Rebecca, 22, 121
Solutions (publication), 70–71
Spark, Amy, 21
spheres of influence, 57–58
Stoknes, Per Espen, 36–38, 72, 102
Stone, I. F., 59
storytelling, 80–81, 86–89, 92–93. *See also* communication
straight-line instinct, 85
stress, 4, 45, 59, 75, 119. *See also* mental health
Student Leadership Institute for Climate Resilience (SLICR), 92, 166n
student strikes, 8, 56, 106
suffering, 42, 44–46, 123, 128–30
suicide and suicidal ideation, 18, 20, 24–25. *See also* depression; mental health
Sunrise Movement, 6, 23–24
survivance, 139–40
Suzuki, Shunryu, 43
system change *vs.* social change, 25–27

Founded in 1893,
UNIVERSITY OF CALIFORNIA PRESS
publishes bold, progressive books and journals
on topics in the arts, humanities, social sciences,
and natural sciences—with a focus on social
justice issues—that inspire thought and action
among readers worldwide.

The UC PRESS FOUNDATION
raises funds to uphold the press's vital role
as an independent, nonprofit publisher, and
receives philanthropic support from a wide
range of individuals and institutions—and from
committed readers like you. To learn more, visit
ucpress.edu/supportus.